Y0-BCC-653

TIME AND AGE

Time Machines, Relativity and Fossils

DB
.X
.W64
2015

TIME AND AGE

Time Machines, Relativity and Fossils

Michael Mark Woolfson

University of York, UK

ICP   Imperial College Press

Published by

Imperial College Press
57 Shelton Street
Covent Garden
London WC2H 9HE

Distributed by

World Scientific Publishing Co. Pte. Ltd.
5 Toh Tuck Link, Singapore 596224
USA office: 27 Warren Street, Suite 401-402, Hackensack, NJ 07601
UK office: 57 Shelton Street, Covent Garden, London WC2H 9HE

Library of Congress Cataloging-in-Publication Data
Woolfson, Michael M. (Michael Mark)
  Time and age : time machines, relativity, and fossils / Michael Mark Woolfson, University of York, UK.
    pages cm
  Includes bibliographical references and index.
    ISBN 978-1-78326-583-1 (hardcover : alk. paper) -- ISBN 978-1-78326-584-8 (pbk. : alk. paper)
  1. Time measurements. 2. Clocks and watches--History. 3. Space and time. 4. Special relativity
(Physics) 5. Statistical astronomy. 6. Earth (Planet)--Age. 7. Geological time. I. Title.
  QB213.W84 2015
  529--dc23

                                                        2014047865

British Library Cataloguing-in-Publication Data
A catalogue record for this book is available from the British Library.

Copyright © 2015 by Imperial College Press

All rights reserved. This book, or parts thereof, may not be reproduced in any form or by any means, electronic or mechanical, including photocopying, recording or any information storage and retrieval system now known or to be invented, without written permission from the Publisher.

For photocopying of material in this volume, please pay a copying fee through the Copyright Clearance Center, Inc., 222 Rosewood Drive, Danvers, MA 01923, USA. In this case permission to photocopy is not required from the publisher.

Typeset by Stallion Press
Email: enquiries@stallionpress.com

Printed in Singapore

Introduction

The concept of time is one that is familiar to everyone, yet it is difficult to define. We know that it can be categorized as past, present and future, the present being defined by events in the process of taking place, the past by events that have already taken place — in principle something that is known — and the future by events that will take place — which are unknown, although we may make predictions about what is expected to happen. The passage of time is marked by events and events are what create change; any event that takes place changes *what will be* to *what is* at the time it happens. We mark the passage of time by change — the movement of the Sun in the heavens or the hands of a watch — and it has been argued that without change, time would not exist.

The idea of time travel has been advanced in several works of fiction. In *The Time Machine*, written by H.G. Wells and published in 1895, an English inventor creates a machine that transports the passenger through time. He travels to a time more than 800,000 years in the future inhabited by child-like creatures, the Eloi, who live an idyllic life of idleness, and a savage ape-like troglodyte race, the Morlocks, who live underground and whose main diet is the Eloi. More recently, in 2006, BBC television produced a series, *Life on Mars*, chronicling the story of Sam Tyler, a detective with Greater Manchester Police who is seriously injured in a car crash and is undergoing drastic surgery. While under anaesthetic, he wakes up in 1973 as a detective in Manchester. He finds himself in a world he struggles to understand. His superior, Gene Hunt, is a tough, ruthless policeman with no regard for political correctness who, when Sam reminds him about human rights when he maltreats a vicious criminal, replies, 'Human rights are for human beings.' These time-travelling narratives are indeed pure fantasy — Sam Tyler even sees his mother as a woman of about his own age accompanied by himself as a child — but they make for good entertainment.

These are philosophical and fantastical observations about the nature of time, but here we shall be concerned with more mundane and realistic matters. The way that time has been, and is, measured will be described both from the point of view of day-to-day life and with the high precision demanded by modern science. When Einstein introduced his Theory of Special Relativity in 1905, time was presented in a new way as intimately connected to space, the two being bundled together in a concept known as *spacetime*. There is a chapter devoted to the consequences of this theory, many of them inconsistent with our instinctive understanding of the way the world behaves and seemingly fantastic — but these consequences can be scientifically shown to be true. The final section of the book is devoted to finding the ages of various kinds of object, from the age of the Universe to the age of a mediaeval artefact and everything between.

The book is intended for the non-specialist reader interested in science in a general way. As far as possible, mathematical treatments should be avoided in such a book, but I have included some mathematics — not much — for those that can deal with it. However, I have also ensured that there is a verbal explanation of the outcome of the mathematical treatments, backed up by diagrams where possible. Have I succeeded in my objective of presenting a book for all the target readers? Only time will tell!

WITHDRAWN

Contents

Time and Relativity

The Ages of Astronomical Entities

The Measurement of Time

Chapter 1

Astronomical Time

1.1 Early Concepts of Time

Early hominids would have had the concept of time well before our own species, *Homo sapiens,* came into being. The daylight period was the time for maximum activity when vision came into play for hunting, the gathering of food and the making of artefacts. The darkness of the night was relieved by the arrival of the period around the full Moon, although once the art of fire-making was mastered a new source of light was available, which, together with toolmaking, marked the beginning of the development of technology. For those well away from tropical regions the progress through the seasons — spring, summer, autumn and winter — would, again, be markers of the position within a yearly cycle.

There would also have been appreciation of the progress of time within these basic time units. The daylight period reaches its peak when the Sun is highest in the sky and morning and afternoon would have become time concepts together with the onset of sunrise and sunset. From the appearance of the new Moon the crescent grows, reaches a peak at full Moon and then declines back to new Moon again. The position within the lunar month would be an obvious way of marking the passage of time. These crude units of time, and the progressions within them, were as obvious in those early days as they are now, but in these scientific times we know much more about them — how they come about and why they are actually variable entities and not the absolute measures of time that they seem to be.

1.2 The Day

Most people know that the length of the day is governed by the rate at which the Earth spins on its axis and many will assume that this is at one revolution every 24 hours — but actually that is not true. The period of its spin is about 23 hours 56 minutes and the 24 hour period between successive occurrences of midday is due to the combined effect of the Earth spinning about its polar axis and the motion of the Earth around the Sun. This is illustrated in Figure 1.1 where, for clarity, the motion of the Earth around the Sun in one day has been exaggerated. Point O on the Earth corresponds to midday, when the Sun is highest in the sky, at Earth positions E_1 and E_2 and the time between these two positions corresponds to a period of 24 hours. However, the figure shows that between the two positions the Earth has spun by more than one complete revolution on its axis; the extra angle through which it has to spin is 1/365 of a complete revolution. Hence the time for one complete revolution is $24 \times 365/366$ hours, which is the spin period of 23 hours 56 minutes.

A further complication in defining the day is that the angular speed of the Earth in its orbit around the Sun varies throughout the year. This is because the Earth's orbit is not a circle but an ellipse and the

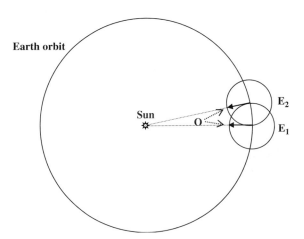

Figure 1.1   The Earth at two positions corresponding to midday at point O (not to scale).

distance of the Earth from the Sun varies from 1.521×10^8 km to 1.471×10^8 km. The Earth has a greater angular speed when it is closer to the Sun but its axial spin is at a constant angular speed. The effect of this is that the period from one midday to the next is longer when the Earth is closer to the Sun; the 24-hour day, as defined by the period between two successive occurrences of midday, is an average and varies throughout the year.

Another complication is that the day is slowly but steadily lengthening. This is due to the tidal interaction of the Earth and the Moon. The day is shorter than the month, so the Earth's spin drags the tides forward making the high tide direction ahead of the direction of the Moon (Figure 1.2). There are two positions of high tide on opposite sides of the Earth, but the effect of the one nearer to the Moon is dominant and it exerts a force on the Moon that has a component in its direction of motion, thus increasing the energy of the Moon in its orbit. This energy, plus the loss of energy due to the friction of the tides being dragged over the Earth's surface — likened to the dragging of brake pads on a spinning wheel that turns mechanical energy into heat — both come from the slowing down of the Earth's spin. There is an additional tidal effect due to the Sun and the total combined effect, due to the Moon and the Sun, is that the length of the day increases by 1.6 milliseconds[1] per century. While this may seem a trivial change it has noticeable consequences in the span of human history. The Babylonians kept records of when eclipses of the Sun occurred and the times when they took place can be calculated. Based on the length

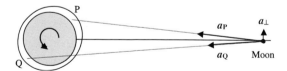

Figure 1.2   The acceleration of the Moon due to the nearside bulge, a_P, and the far-side bulge, a_Q, with a component $a_\perp$ in the direction of the Moon's motion.

---

[1]A millisecond (ms) is 10^{-3} s, one-thousandth of a second.

of the day being constant at its current value, such calculations indicate that the Babylonians were observing eclipses of the Sun in the middle of the night! If the day has lengthened over the last 4,000 years at a rate of 1.6 ms per century, then the length of the day has changed by 40×1.6 ms $= 64$ ms or, over the 4,000-year period, it has been on average 32 ms shorter than now. Thus the time slippage over 4,000 years is $0.032 \times 4000 \times 365$ s $= 13$ hours, a value that explains the ancient records. Although this rate of change is small, it means that 200 million years ago, when dinosaurs first appeared on Earth, the length of the day would have been just over 23 hours, taking an hour as presently defined.

The scientific unit of time is the second, which was traditionally defined in terms of the average day, taken to consist of 24 hours each divided into 3,600 seconds. It is clearly unsatisfactory to have a scientific unit defined in terms of a quantity that is variable, so now the second is defined in another way that is invariant and is described in Section 4.3.1.

Apart from the systematic variations in the length of the day — throughout the year due to the Earth's orbit and over time due to Earth–Moon and Earth–Sun interactions — there are also small irregular fluctuations due to changes either in the distribution of the Earth's atmosphere or of matter in the Earth's interior. However, although of interest scientifically, they are far too small to be detected, except by the most refined measurements.

1.3 The Month

There are two ways of defining the period of the Moon's orbit around the Earth. The *lunar sidereal period* (27.322 days) is the time taken for the Earth–Moon direction to go through one revolution, i.e. an angle 2π in radians,[2] relative to the fixed stars. The *lunar synodic period* (29.531 days), which we normally call the *lunar month*, is the time taken to go from one full Moon to the next. The sidereal period corresponds to one complete orbit around the Earth. In Figure 1.3

---

[2] A radian is the angle subtended at the centre of a circle by an arc equal in length to the radius. One revolution corresponds to an angle 2π in radians.

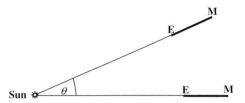

Figure 1.3 The Earth (E), Sun and Moon (M) at the beginning and end of a synodic period.

the Sun–Earth–Moon configurations are shown at the beginning and end of a synodic period, t_{syn}. Since the Earth has moved round the Sun at an angle θ, it follows that

$$t_{syn} = \frac{\theta}{2\pi} t_y \tag{1.1}$$

where t_y is a period of one year. The synodic period corresponds to a rotation of $2\pi + \theta$ around the Earth while the sidereal period corresponds to a rotation of 2π so that

$$\frac{t_{syn}}{t_{sid}} = \frac{2\pi + \theta}{2\pi}. \tag{1.2}$$

Eliminating θ from (1.1) and (1.2) gives

$$\frac{1}{t_{syn}} = \frac{1}{t_{sid}} - \frac{1}{t_y}, \tag{1.3}$$

which is now confirmed numerically, given $t_y = 365.242$ days.

$$\frac{1}{t_{syn}} = \frac{1}{\text{lunar synodic period}} = \frac{1}{29.531\,\text{days}} = 0.03386 \text{ days}^{-1},$$

$$\frac{1}{t_{sid}} - \frac{1}{t_y} = \frac{1}{\text{lunar sidereal period}} - \frac{1}{\text{year}}$$

$$= \frac{1}{27.322 \text{ days}} - \frac{1}{365.242 \text{ days}} = 0.03386 \text{ days}^{-1}.$$

Once again, the definition given is for a *mean* synodic month since the Moon's orbit around the Earth is an ellipse, with maximum and minimum distances from the Earth 405,900 and 362,900 km respectively.

The Earth–Moon tidal interaction that gives a constant reduction in the rate of the Earth's spin also increases the energy of the Moon's orbit, which means that it is gradually retreating from the Earth. This is at a rate of 2.9 cm per year which, although slow, is quite significant over a long period of time. For example, in 1,000 million years' time, the Moon will have retreated by 29,000 km; its angular diameter will be reduced by 7% and a total eclipse of the Sun will be impossible. At that time the lunar sidereal period will be about 30.6 days.

1.4 The Year

A year is the time taken by the Earth to make one complete orbit of the Sun. That simple statement defines the year precisely — 365.2422 days — and is virtually free of complication except that the length of a day is slowly varying. The gravitational attractions of other planets of the Solar System do slightly affect the Earth's orbit over a long period, thus giving small variations to the year but these are outside the range of being noticeable on historical timescales.

1.5 The Influence of Time on Life

All life is affected by the passage of time, on a daily, monthly and yearly basis. There is a large variety of reactions for different life forms but here we just give a selection of some of the more important ones.

1.5.1 *Plants and time*

Vegetation is dependent on energy from the Sun to perform photosynthesis and will usually orient its leaves to receive the maximum amount of sunlight. Many flowers close their petals at night or in inclement weather to protect the stamen and pistil, the reproductive organs, and to reduce moisture loss by respiration. These behaviour patterns come about through chemical changes and do not involve the existence of the 24-hour day as a unit of time. It is possible to grow plants under artificial light, to which they will react as though it were sunlight, and so to vary the apparent length of the day, to which they will react quite normally.

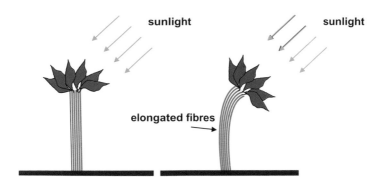

Figure 1.4 The bending of a plant towards light.

Obtaining maximum exposure of its leaves to the Sun's rays is advantageous to the plant because the action of sunlight on chlorophyll, the green pigment found in leaves, enables a chemical reaction between water drawn up by the plant's roots and atmospheric carbon dioxide, to give cellulose, which becomes the structural material of the plant, and oxygen. Charles Darwin wrote the book *The Power of Movement in Plants* about this phenomenon of *phototropism*, movement in plants. We now know that this motion is caused by chemical materials, one of which is *indole-3-acetic acid* (IAA), one of a class of plant hormones called *auxins*. Light falling on the plant from one direction causes the IAA on the other side of the plant to elongate the fibres in its vicinity, which bends the plant towards the light (Figure 1.4).

There are different opinions about whether or not plant life reacts to the phases of the Moon but of its reaction to a yearly cycle there is no doubt. As the seasons change, so do the hours of daylight and the mean temperature, and many trees, shrubs and flowers essentially hibernate during the cold parts of the year until the return of warmth and sunlight enables them to grow again and to propagate through the production of seeds or spores — or in other ways such as the production of underground runners from which new plants grow. Seasonal effects on plants are strongest in regions remote from the tropics, since the variation in the length of day and temperature are greater. In many tropical, or sub-tropical, regions, where variations in temperature and daylight are less, it is possible to grow two crops per year. However, in some tropical regions there are seasonable variations in plant life due to periods of heavy rainfall, such as the *monsoon* season of south-east Asia. These are caused by variations of sea temperature and winds, reacting with topographical features, such as the high Tibetan Plateau.

1.5.2 *Animals and time*

Animal reaction to time, as expressed by the occurrence of days, months and years is much more complex than that of vegetation. The majority of predators hunt during the day when they can see their prey and then either ambush them or run them down as they attempt to escape. Other predators operate at night, which can be because their prey is only available at that time, because they themselves are less likely to fall victim to other predators or because the competition from other predators is less. For example, many bats — those termed *microbats* that eat insects — are usually nocturnal hunters because many of the insects they eat, such as moths, are themselves nocturnal. Evolution has equipped these bats with echolocation abilities that enable them to detect and home in on their prey even in complete darkness.

Animal reactions to the lunar cycle are usually influenced by the presence of light during full-Moon periods. However, the Moon does more than just give light during the night — it also has a strong influence on tides. At times of either the full Moon or the new Moon, the Sun and Moon constructively combine their tidal effects to give very high tides — called *spring tides* (the term 'spring' comes from the German *springen*, meaning 'to rise up', and has nothing to do with the time of year). During early summer, at the times of spring tides, some types of crab, notably the *horseshoe crab*, invade sandy beaches to lay their eggs. Presumably, this has the advantage that firstly they have sufficient light to produce the nest for laying their eggs and secondly they have less far to go up the beach to lay eggs in places that will not be inundated, with consequential risk to the unhatched eggs.

Just as plant life, animals are affected by the yearly cycle, although in many different ways. Animals that live in tropical or sub-tropical regions of the Earth normally carry on with their lives in an unchanged way throughout the year. By contrast, animals that live in northern temperate or polar regions need to adapt their lifestyles, and sometimes even their locations, to cope with changing conditions. At one extreme, some creatures, such as bears, bats and rodents, behave like some plants and simply shut down activity during the cold season by hibernation. The tactic is to a accumulate a considerable store of fat in the period preceding hibernation and then, in a sheltered environment which could be a burrow or a cave, to go into a state of torpor or sleep during the winter months, during which the metabolic rate is slowed down, characterized by a low heart rate, slow breathing and a lowering of the body temperature.

Global warming, whatever its cause, is having a serious effect on many animal species. Polar bears need sea ice to provide a platform for hunting seals and, as the years go by, there is less sea ice with greater distances between ice sheets. It is getting increasingly difficult for polar bears to build up sufficient stores of fat to carry them over the summer period when hunting opportunities are few, or even non-existent.

Other creatures meet their changing needs during the year by migration. The two driving forces for migration are the availability of food and the need to breed. In polar regions there is ample supply of food in the cool seas, particularly krill, an abundant crustacean, but also many kinds of fish. Several species of whales spend the summer period in the prolific feeding grounds of either the Arctic or Antarctic and then move south or north to warmer climes for breeding in the winter season. There are also many migrating species of birds. The champion of them all is the *Arctic tern*, which travels about 70,000 km each year, breeding in Arctic regions in the northern summer and then moving to the Antarctic to benefit from the greater food supply there in the southern summer.

1.5.3 *Humans and time*

Of all the animals, mankind alone has been able to overcome the difficulties of the annual variations of temperature and food supply to be able to live almost everywhere on Earth, excepting in extreme polar regions and in some exceptionally dry desert areas. The ability to make clothing and build shelters has enabled mankind to live in conditions of extreme cold and their ingenuity in making weapons enabled early human species to hunt and kill animals much stronger and faster than themselves, ensuring the survival of the species.

Again, alone of all animals, mankind has been able to understand the nature of the daily, monthly and yearly changes in conditions. The development of civilized societies, dependent on agriculture and artisan skills, brought in new needs to measure time. The position and height of the Sun provided a rough-and-ready measure of the passage of time during the day, but it was also required to know the time of year when crops should be planted. The development of religion, and religious festivals, also gave the incentive to construct a calendar of some kind and it was this that provided a strong motivation for astronomical studies.

1.6 Calendars

The Sun and the Moon move in a regular pattern across the sky and the lunar month and the year became periods of time that could be defined and found with certainty and precision, even by early societies. The earliest calendars, designating the time of year, were based on the lunar month, but since the month — the lunar synodic period — was not a whole number of days and the year was not a whole number of lunar months, various ways had to be devised to keep the calendar in step with the time of year. Here we describe a selection of calendars that illustrate the various schemes that have been used.

1.6.1 *The Sumerian calendar*

Sumer was an ancient civilization, founded about 4000 BCE, situated in the area of what is now Iraq. The Sumerian city of Ur was the birthplace of the biblical character Abraham sometime in the second millennium BCE. The Sumerian calendar was divided into 12 lunar months, each of which began at the time of the new Moon and hence consisted of either 29 or 30 days. This gave an average year of $12 \times 29.531 = 354$ days to the nearest whole number. To keep their calendar roughly in step with the solar year, every three years they inserted an extra month. This type of calendar, based on 12 lunar months with an extra month added periodically to synchronize with the solar year, was also used in ancient Greece and is the basis of the modern Hebrew calendar. On the other hand, the Islamic calendar retains 12 lunar months without the periodical addition of an extra month, so the Islamic year moves by about 11 days every year with respect to the solar year.

1.6.2 *The Mayan calendar*

The Maya civilization thrived in Central America from about 1800 BCE until about AD 900, from which date it fell into a state of decline. It was finally extinguished by the invading Spanish conquistadors in the 16th century. At the height of their power the Mayans exhibited knowledge of astronomy far better than that of contemporary Europe and their estimate of the length of the year (365.242036 days) was surprisingly accurate, especially given the time it was made.

The Mayans had two quite independent calendars, one, called the *Haab*, based on a 365-day cycle, and the other, the *Tzolkin*, a 260-day year used for religious and spiritual matters, such as predicting the future or finding the most auspicious dates for various activities, e.g. making war. The Haab was fairly straightforward with 20 months, each of 18 days. The 5 extra days to make the total number of days in the year 365, were called *Uayeb* and were associated with unpleasantness, such as bad luck, disease and death. According to Mayan beliefs, anyone born in the Uayeb period would have an unhappy and disastrous life.

The Tzolkin year was much more complicated. There were 20 different day names, shown in Figure 1.5, and 13 numbers. The first day of the Tzolkin cycle was 1-Imix, the second 2-Ik, the third 3-Akbal, continuing as shown by the vertical arrows in the figure. After 260 days, 1-Imix would occur again and the cycle was complete. A festival fixed on a certain day in this calendar would wander through the solar year but in an equatorial zone there is little feeling of association between festivals and weather. By contrast, people in Canada would find it odd if Christmas were a day of high temperature and those in Australia would be equally astonished if Christmas day were frosty; the Mayans would not be so conscious of the 105-day shift in the solar year between successive occurrences of the same festival.

Numbers

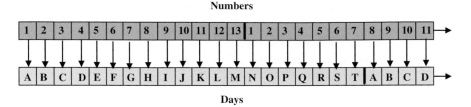

Days

Figure 1.5   The beginning of the Mayan Tzolkin calendar period. The day names are:

A = Imix	B = Ik	C = Akbal	D = Kan	E = Chiccan
F = Cimi	G = Manik	H = Lamat	I = Muluc	J = Oc
K = Chuen	L = Eb	M = Ben	N = Ix	O = Men
P = Cib	Q = Caban	R = Eiznab	S = Cauac	T = Ahau

The combination of a 365-day Haab and a 260-day Tzolkin meant that every 18,980 days (the lowest common denominator of 260 and 365), or 52 years, the two calendar systems returned to the same relative state and this period was referred to as a *bundle* and played the role in Mayan society that the century plays in our own.

1.6.3 *The Roman, Julian and Gregorian calendars*

The beginning of the modern western calendar is said to derive from the founding of Rome in 753 BCE with what is called the *Romulus calendar*. This contained just ten months, listed below with the number of days in parentheses:

Martius (31)	Aprilis (30)	Maius (31)	Iunius (30)
Quintilis (31)	Sextilis (30)	Septembris (30)	Octobris (31)
Novembris (30)	Decembris (30)		

It will be seen that this year contains only 304 days; the winter period from the end of Decembris to the beginning of Martius was not assigned to any month. The origins of the modern names for months will be seen in some of the above. In particular, the final six are related to the Latin names for five (quinque), six (sex), seven (septem), eight (octo), nine (novem) and ten (decem).

This rather unsatisfactory calendar, leaving part of the year unassigned, was refined by an early Roman king, Numa Pompilius (753–673

BCE), in 713 BCE. He prefixed the ten months of the Romulus calendar with two extra months, *Ianuarius* and *Februarius*, with 29 and 28 days respectively. This had the effect of making the names of the final months less logical, e.g. Quintilis was now the seventh month, and it still fell short of the full number of days for a solar year. To compensate for this last point, from time to time a leap month, *mensis intercalaris*, was inserted in the year to bring the calendar back into line with the solar year.

The next change, made at the time of Julius Caesar (100–44 BCE) in 46 BCE, was to alter the number of days in each month to give 365 in total, thus giving rise to the *Julian calendar*. This was further refined in 44 BCE by his successor Augustus (63 BCE–AD 14) who created the 365.25-day calendar year by instituting the extra day in Februarius every four years. Augustus also changed the name of the month Quintilis to *Iulius* in honour of his predecessor and, finally, in 8 BCE sextilis was renamed *Augustus*. The resultant Julian calendar, with names modified to fit different language requirements, is the one that we recognize and hang on our walls today.

The difference between the Julian calendar year, 365.25 days, and the solar year, 365.2422 days, meant that there was slippage between the two. The Roman Catholic church was particularly concerned because the date of Easter, traditionally a spring festival, was gradually moving towards the summer. A German Jesuit priest, Christopher Clavius (1536–1612), found a solution to this problem which was ordained by Pope Gregory XIII (1502–1585) in 1582 and became known as the *Gregorian calendar*. The day after October 4th 1582 became October 15th, which restored Easter to its proper place in the solar year, and the pattern of leap years was changed to bring the calendar more into line with the solar year. Under the new system, a leap day, February 29th, was introduced every four years (as before) with the exception that there was no leap year in centenary years (years divisible by 100) unless the year was divisible by 400, e.g. the year 2000 *was* a leap year but the year 1900 was not. This changeover to the Gregorian calendar was not universally accepted when it was proposed. Dominantly Catholic countries like Spain, Portugal and Italy adopted it almost immediately, but other countries followed suit rather slowly — Britain and its Empire in 1752 and Russia and other Eastern European countries in the early part of the 20th century.

1.7 The Changing of the Seasons

The Gregorian calendar is almost, but not quite, in line with the solar year, since it gives a year of 365.2419 days. However, the difference is small and there are other, more effective, processes at work that lead to a drift of the seasons with respect to the calendar over long periods of time. One of these is the change in the length of the day; we saw in Section 1.2 how this explains Babylonian observations of eclipses of the Sun. There is also another effect, which is of much greater importance.

The cause of seasonal changes during the year is the 23½° tilt of the Earth's spin axis with respect to the *normal* (the perpendicular) to the *ecliptic*, the plane of the Earth's orbit around the Sun. In the northern summer, the northern hemisphere is tilted towards the Sun and six months later, when the Earth has moved to the diametrically opposite point of its orbit, it is the southern hemisphere that points towards the Sun and gives the southern summer (Figure 1.6). This effect takes place because the direction of the Earth's spin axis is fixed in relation to the stars, but over long periods of time the axis changes direction. This is due to the gravitational attraction of the Moon on the tides it raises on Earth, which causes the spin axis to undergo *precession* with a period

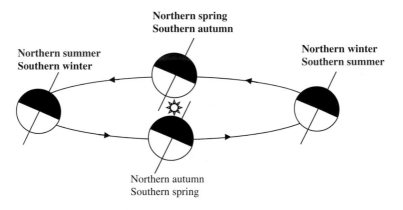

Figure 1.6   The variation of the seasons as the Earth moves round the Sun.

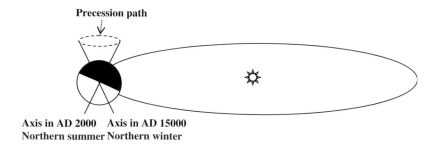

Precession path

Axis in AD 2000   **Axis in AD 15000**
Northern summer Northern winter

Figure 1.7   The effect of precession of the Earth's spin axis after 13,000 years.

of 26,000 years. This kind of motion is illustrated in Figure 1.7. What this means is that 13,000 years in the future the time of the northern summer solstice, when the Sun is at its most northerly latitude, will have moved from its present date of about 21st June to about 21st December. If humankind as a species is still around in thousands of years from now, it will either have to accept the drift of the seasons or undertake other periodic modifications to the calendar.

Chapter 2

Early Recording of Time

2.1 Sundials

As a consequence of the Earth's spin, the Sun is seen to rise in the east, climb higher above the horizon, reach its zenith at midday and then sink down, eventually to set in the west. Thus the position of the Sun in the sky can be used as a marker for the passage of time during daylight hours but, clearly, directly measuring the position of the Sun is a difficult thing to do. It was noted by early societies that the passage of the Sun could more conveniently be followed by the motion of the shadows it produced and this led to the invention of the *sundial* as an instrument for determining time during the day. There is definite archaeological evidence for the use of sundials in Egypt, and probably in other parts of the Middle East, in about 1500 BCE.

In its simplest form, a sundial consists of a thin rod, or some object with a straight edge, called a *gnomon*, the shadow of which is thrown onto a surface on which hour lines are marked to indicate the time. There is a great variety of sundial designs, varying in the orientation of the gnomon and the type and orientation of the surface on which the shadow falls. Here we shall concentrate on a simple, and most common, form of sundial and briefly indicate some of the other forms that have been made.

2.1.1 *The equatorial sundial*

In considering the action of a sundial, it simplifies matters if the Earth is considered as not spinning and the Sun orbiting the Earth within a period of one day. The most straightforward sundial is when the

19

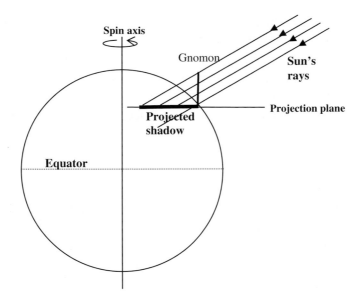

Figure 2.1   A sundial with the gnomon parallel to the Earth's spin axis and the projection plane parallel to the equator.

gnomon is parallel to the Earth's spin axis, for then its orientation remains constant. Now, as the Sun apparently orbits the Earth, the shadow of the gnomon will rotate at a steady rate, at 15° per hour corresponding to a 360° revolution in 24 hours. If the shadow is thrown onto a plane that is perpendicular to the gnomon, so that it is parallel to the Earth's equatorial plane, then the hour marks on the plane will be a set of lines, radiating outwards from the gnomon, at 15° intervals. This arrangement, known as an *equatorial sundial*, is shown in Figure 2.1. If the sundial location is at latitude λ then the gnomon must point at an angle 90° — λ to the vertical and be directed along the local meridian.

From the point of view of the Earth, the rotation of the Sun is at different latitudes at different times of the year — at 23½° north at the northern summer solstice, in the equatorial plane at both the vernal (spring) and autumnal equinox and at 23½° south at the northern winter solstice. From Figure 2.1 it will be seen that if the sundial is in the northern hemisphere, no shadow will be cast by the gnomon in the winter, although the Sun will still be above the horizon and be capable of casting shadows. The answer to this problem is to have a different arrangement to record the shadow. One solution is shown in Figure 2.2, which shows a large equatorial sundial situated in the Forbidden City in Beijing. There is a gnomon on both sides of the projection plane so that between the autumnal and vernal equinoxes the shadow is thrown onto the lower surface. The design of a portable equatorial sundial, which operates at all times of the year, is shown in Figure 2.3. The angle of the gnomon can be adjusted for latitude and its shadow is projected onto the part-circular strip on which the hours are marked in equal 15° intervals. For the proper recording of time the vertical plane containing the gnomon must be along the meridian.

For an equatorial sundial, such as that shown in Figure 2.2, the length of the shadow of the gnomon throughout any day will remain approximately constant, ignoring the very small change in the latitude of the apparent orbit of the Sun during a single day. This means that the shadow of the tip of the gnomon, called the *nodus*, follows an arc of a circle on the projection plane. However, the length of the shadow depends on the Sun's orbital latitude that, in its turn, depends on the date. Figure 2.2 shows date lines as concentric circles marked on the lower projection plane. These give the date, with limited precision, between the autumnal and vernal equinoxes and circles on the upper projection plane give approximate dates between the vernal and autumnal equinoxes.

A correction that must be made to sundial time relates to its position in the time zone within which it operates. Greenwich Mean Time (GMT) is used in the United Kingdom, which is the time based on midday defined by when the Sun is at its zenith in Greenwich, just outside London. For those living in Cornwall, some 400 km to the west of London, the Sun reaches its zenith about 22 minutes later. For a fixed sundial, this is corrected by rotation of the hour lines inscribed on the projection plane; for portable sundials the correction involves either rotation, so that the gnomon is not directed along the meridian, or just knowing what correction must be applied. Of course, in ancient civilizations time zones did not exist and all time was measured on a local basis; indeed the UK operated on local time until 1840, when the

Figure 2.2   An equatorial sundial in the Forbidden City, Beijing (photo credited to Sputnikcccp).

Figure 2.3   A year-round portable equatorial sundial.

necessity of having a standard time for railway timetables established the present system.

2.1.2 *The equation of time*

In the description we have given of the equatorial sundial, it has been assumed that the Earth's orbit, or the apparent orbit of the Sun, is circular but it is actually an ellipse. The effect of this was explained in Section 1.2; when the Earth is near *perihelion*,[1] the time from one solar zenith to the next is longer than average and near *aphelion*[2] it is shorter. We now measure time by devices that make every day the same mean length and so there is a gradual drift in the time recorded on a sundial to that referred to in the UK as GMT. Near perihelion the sundial is losing time and the accumulated loss of time can reach about 17 minutes. Near aphelion the situation is reversed and eventually the sundial time moves ahead of GMT and the correction needed is to subtract several minutes.

There is another contribution to this drift of sundial time away from GMT, which is the inclination of the apparent orbit of the Sun, due to the 23½° tilt of the Earth's spin axis (Figure 1.6). During the day, as the Earth spins on its axis, it also moves round in its orbit. This affects the motion of the shadow so it will not move at precisely 15° per hour as it would if the Earth were stationary. However, the component in

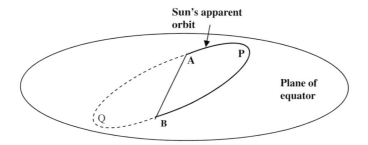

Figure 2.4 The Sun's apparent motion relative to the plane of the Earth's equator.

---

[1] The perihelion is the point on a planetary orbit when it is closest to the Sun.
[2] The aphelion is the point on a planetary orbit that is furthest from the Sun.

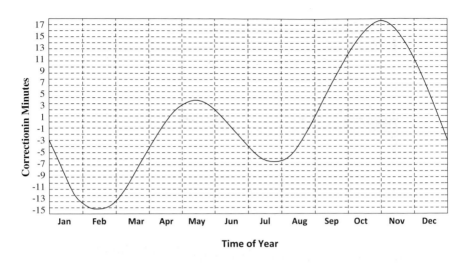

Figure 2.5   The equation of time for corrections to sundial time.

the equatorial plane of the Earth of the speed of the Earth round the Sun varies throughout the year. In Figure 2.4, we show this by looking at the apparent motion of the Sun relative to the Earth, assuming a constant speed in a circular orbit. At points P and Q the motion is parallel to the equator and the component in the equatorial plane is the orbital speed; at A and B the motion is inclined to the equatorial plane and so the component of speed in the equatorial plane is less than the orbital speed.

The net result of the effect of the non-circular nature of the Earth's orbit and the tilt of its spin axis is that the correction that has to be made to the time indicated by the sundial to obtain GMT varies throughout the year. This correction, known as the *equation of time*, is shown in Figure 2.5; it is engraved on the base of the sundial shown in Figure 2.3.

2.1.3 *Other designs of sundials*

It is quite common to have a *horizontal sundial* in which the gnomon is aligned with the Earth's spin axis but the projection plane is horizontal. This has the advantage that it operates year-round, since during daylight the Sun is always able to cast a shadow of the gnomon. The disadvantage is that the hour lines are no longer uniformly spaced in angle. This can

Figure 2.6   The horizontal sundial outside the Supreme Court in Perth, Western Australia.

be seen in Figure 2.6, which shows a sundial outside the Supreme Court in Perth, Western Australia.

Another common arrangement, especially on the sides of buildings, is where the projection plane is vertical. Clearly, the wall on which the sundial is mounted must be south facing (in the northern hemisphere) and in the summer it will only be able to record time for a period of 12 hours, although days may be much longer. A typical *vertical sundial* is shown in Figure 2.7; again the non-uniform spacing of the hour lines is a feature of this design.

Although sundials have been used from antiquity, their disadvantages are evident — they depend on the presence of sunlight, which in many parts of the world is intermittent and in all parts of the world is non-existent during the night. For this reason people in ancient civilizations found other ways to record time that could be used anywhere and at any time.

2.2 Candle Clocks

The whole basis of a device for measuring time is that something should move at a predictable, preferably uniform, rate and the position reached

Figure 2.7   A vertical sundial at Moot Hall, Aldeburgh, Suffolk, England.

then becomes an indicator of the passage of time. In the case of the sundial, the moving entity is a shadow that moves at a more or less steady rate for an equatorial sundial but at a predictable rate for other sundial arrangements.

The earliest form of lighting apart from fire, whose main purpose was to provide warmth and a means of cooking, was the oil lamp. They gradually improved and in the 7th century BCE Greeks were making terracotta oil lamps with wicks. The next development in lighting technology was the invention of the candle, probably by the Chinese in the 3rd century BCE. They initially used whale oil as the source of wax, but later in China, India and Japan other sources were found — for example from insects, boiling cinnamon, nuts and beeswax. The Chinese moulded their candles in paper tubes and used rice paper as a wick.

Exactly when the idea arose of using a candle as a time-recording device is not known, but in AD 520 a candle clock was mentioned by

You Jiangu in a Chinese poem. It would have been a simple device — a thin candle of uniform width with equal distances, giving equal intervals of time, marked along its length, placed in a protective housing to shield it from draughts.

Eventually, the art of candle making and the idea of using candles as clocks reached the Islamic culture, which flourished in the Middle East and surrounding areas. A 12th-century Muslim scholar, Al-Jazari (1136–1206), a brilliant inventor, engineer, skilled craftsman and mathematician, designed a candle clock that recorded time with a dial. This was described in his *Book of Knowledge of Ingenious Mechanical Devices*, which contained descriptions of 100 mechanical devices of his own invention. The candle clock, the book illustration of which is shown in Figure 2.8, has the top of the candle pushed upwards against a round metal plate with a hole at the centre through which

Figure 2.8 Al-Jazari's recording candle clock.

the wick protrudes. The candle is pushed upwards by a baseplate that has upward pressure applied to it by a pulley-and-weight arrangement. As the candle burns away, the baseplate moves upwards and this motion is converted into the movement of a dial that records the time.

The art of using candles as clocks eventually penetrated to Western Europe. The Saxon king Alfred the Great (849–899) was said to have used a candle clock consisting of six candles, consecutively burning for four hours each, so covering a 24-hour day, with 12 marks on each candle corresponding to 20 minute intervals. The candles were housed in a shielded enclosure to ensure that they burned at a steady rate.

While candle clocks could be used for recording time within a day and could also operate at night when sundials were inoperative, they could not be used to keep track of time over long periods; it would be necessary to establish a new time baseline at intervals using a more reliable method of time determination such as a sundial. Other drawbacks of candle clocks were that they required constant attention to maintain time recording over a long period and also that it was difficult to ensure that they burned at a steady rate.

2.3 Water Clocks

Anyone who has seen a tap emitting a fine stream of water will have noticed its smoothness and uniformity. There is a constant pressure provided by the water main and, because the tap is not completely shut off, there is a small channel through which the water can escape. Of course, there were no water mains and taps in ancient civilizations but, nevertheless, the regularity of water flow through a small channel would have been noted and the idea that this could provide a mechanism for recording the passage of time would have given rise to the *water clock*.

We shall consider three categories of water clock:

1. The *outflow water clock*.
2. The *inflow water clock*.
3. Water clocks incorporating or controlling mechanical systems.

The earliest forms of water clock were rather crude indicators of time and were often calibrated using a sundial. Their operation was also very variable, depending on the prevailing temperature and humidity. They cannot be used at temperatures so low that water would freeze, but even at higher temperatures there are serious problems. The viscosity of water, the property that controls the ease with which it flows, is highly temperature-dependent and this would control the rate at which a stream of water would escape through a narrow channel. Another factor is humidity which, together with temperature, would control the rate at which water evaporates — a serious problem if a quantity of water is being used to measure the passage of time.

2.3.1 *Outflow water clocks*

The *outflow water clock* is a container initially filled with water with a small hole near its base. The falling level of water in the container then indicates the time that has passed since it was full.

When water clocks were first invented and used is uncertain. There are very ancient records, written on clay tablets, which indicate that water clocks were used in Babylon in the period 1600–2000 BCE — the most certain indication of an early use — although some scholars have claimed that water clocks may have originated in China around 4000 BCE.

Babylonian clocks were simple cylindrical containers with a hole at the base. They were not used as clocks in the traditional sense of continuously recording the passage of time, but rather as indicators that a particular period of time had expired. In Babylon, night guards were employed who were paid according to the length of their watches. A night guard would have a longer working time in the winter than in the summer and the times they spent on their work shifts and for which they were paid was controlled by the time it took for a water clock to become empty when initially filled with a certain weight of water. The Babylonian unit of weight was the *mana*, slightly under half a kilogram, and at the summer solstice, when the nights were shortest, a work shift involved emptying two mana of water from the clock. Every two weeks, one-sixth of a mana was added to take account of the lengthening night until the time of the autumn equinox when it became three mana and then, at its peak, four mana by the time of the winter solstice. The same

Figure 2.9   An early Egyptian water clock.

procedure in reverse, removing one-sixth of a mana every two weeks, then returned to two mana at the time of the next summer solstice.

The earliest artefact that can be identified as a water clock comes from a temple at Karnak in Upper Egypt and dates from the reign of Pharaoh Amenhotep III who ruled 1417–1370 BCE. It consists of a vessel with sloping sides and a small hole near the base; time was recorded by the level to which the water had fallen. Figure 2.9 shows the form of the device. As the level of the water fell, so did the pressure pushing the water out of the small hole, so the rate at which water left also fell. By having sloping sides to the vessel, the rate at which the level fell could be made more uniform and the lines on the inside of the vessel, which marked the passage of equal intervals of time, were more equally spaced.

The Egyptians conceived the idea of dividing a 'day' into 12 'hours', but a day was defined as from sunrise to sunset, so the hour — the *temporal hour* — varied throughout the year. Inside the sloping sides of the water clock there were 12 sets of markings, one for each month of the year. Later, the Egyptians divided the day — from one noon to the next — into 24 equal hours. Because the day thus defined varied slightly through the year due to the Earth's elliptic orbit, so did the hour, but the variation was too small to be of concern in a society that could only measure time comparatively crudely.

The ancient Greeks used the name *clepsydra*, literally meaning *water thief* — derived from combining the Greek words for 'to steal' and 'water' — and this word is still in use to refer to a water clock. Both Greeks and Romans used the basic outflow water clock, a simple vessel with a hole near its base, to control the total time allowed for various activities, such as someone presenting their case in court — much as the Babylonians did for timing the watch periods of their night guards.

Still on the theme of limiting the amount of time for an activity to take place, in the 4th century BCE the Persians were using a simple kind of water clock to ration the supply of water to farmers. In Persia, water from distant mountains was directed via underground tunnels, called *qanats*, to semi-arid parts of the country. Farmers were allocated a certain time to take water from a qanat. The timing device consisted of a container of water in which was placed a bowl with a hole at the centre of its base. The bowl gradually filled with water and when it sank it indicated the unit of time. A farmer's ration of water would consist of a fixed number of these time units. This system was used until 1965, when time measurement was replaced by conventional clocks.

2.3.2 *Inflow water clocks*

The disadvantage of the outflow water clock is that the rate at which the water leaves the water container is continuously changing as the water level falls. This requires the use of a containing vessel with sloping sides if the markings on the sides of equal intervals of time are to be more

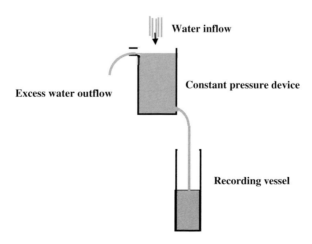

Figure 2.10   A schematic of an inflow water clock.

or less uniform. A schematic design of an *inflow water clock* is shown in Figure 2.10. The water in the vessel providing the steady flow is kept at a constant level by an overflow mechanism so that the pressure providing the flow does not change. This means that the receiving vessel can be of cylindrical shape and equal intervals of water level correspond to equal intervals of time.

The use of outflow water clocks was recorded in China in the 6th century BCE, but from about the 2nd century BCE they were superseded by the inflow type. The problems associated with the changes in the properties of water with temperature and the effects of evaporation were well understood by the Chinese and the performance of water clocks was sometimes checked by comparing them with sundials. In AD 976, Zhang Sixun, an astronomer and military engineer, designed an inflow clock using mercury instead of water, which solved the problem of freezing. In the 14th century, during the Ming Dynasty, the Chinese engineer Zhan Xiyuan used fine sand as the mechanism for a clock; this behaved like a liquid in that it flowed and it eliminated all the problems associated with the changes in the properties of liquids. Indeed, the simple hourglass (Figure 2.11), used in many homes today to time various domestic activities such as boiling eggs, uses sand as a fluid-like material. The hourglass can be rotated to make the top container the one containing the sand and the time taken for it to empty will give the time for the perfect 3-minute egg, or whatever other time the hourglass is designed to deliver.

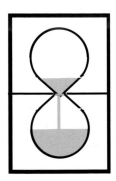

Figure 2.11   A simple hourglass (egg timer).

2.3.3 *Mechanized water clocks*

The Chinese introduced the idea of using a float to activate a pointer moving over a dial to indicate the time. This concept is a familiar part of modern life where a float in a water tank, such as a toilet cistern or the supply tank for a hot-water supply, operates a cut-off mechanism when the water rises to a required level. The attachment of intricate mechanisms to water clocks seems to have been a Chinese speciality. This reached its pinnacle in 1088, when Su Song (1020–1101) constructed a timekeeping device some nine metres tall, operated on the principle of a water clock which incorporated an escapement mechanism (Section 3.2), struck the hours and operated an armillary sphere indicating the position of various astronomical objects in the heavens (Figure 2.12).

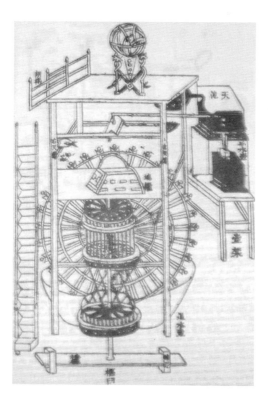

Figure 2.12 Su Song's water clock tower (from Su Song's book describing the tower, 1094).

Figure 2.13   Al-Jazari's castle clock. The birds are perched above the urns into which they drop balls. The musical automata are seen along the bottom and the 12 doors, which display a different figure each hour, can be seen strung out above the archway, with the figure furthest to the left displayed.

The art of making very elaborate water clocks with attached mechanical contraptions thrived in many other societies. The ancient Greeks and Romans were also skilled in the art of constructing water clocks, some of which incorporated dials and pointers to clearly indicate the time and struck bells at regular intervals to mark the passing of hours. Even more elaborate clocks opened flaps to display figures or operated models to show the positions of astronomical objects. However, the most elaborate of all the mechanical water clocks was constructed by Al-Jazari (Section 2.2). In 1206, he constructed his 'castle clock' (Figure 2.13), a veritable tour de force of water clocks, that, in addition to indicating the time, displayed the zodiac showing the motions of the Sun and the Moon, opened doors to show small figures

Figure 2.14   The water-clock sculpture *Man, Time and the Environment*, Hornsby, New South Wales, Australia.

each hour, had two figures of birds that dropped balls into urns and also included five mannequin figures of musicians who played music.

There were other elaborate mechanical water clocks constructed at different times in different places. Even today, designed as artistic curiosities, there are water clocks in various locations, e.g. the Hornsby water clock in Australia (Figure 2.14). However, while water clocks survived until the early mediaeval period they were soon to be replaced by more reliable and potentially more accurate clocks which relied on mechanical devices for recording time rather than the flow of water.

Chapter 3

Mechanical Clocks

3.1 The Salisbury Cathedral Clock

The early water clocks, which recorded time by the amount of water flowing either out of or into a container, gradually become more elaborate and involved mechanical means of indicating the time, as illustrated by Al-Jazari's castle clock (Figure 2.13). From some time in the 13th century, clocks were being designed that operated by purely mechanical means, without the flow of water being involved in the timing process. There are records of early clocks, dating from the late 13th century to the early part of the 14th century, but the oldest clock from just after this early period that is still in working order is the one at Salisbury Cathedral in the UK, which dates from 1386 (Figure 3.1). It was modified in the 17th century to incorporate a pendulum to better control the timing and in 1884 it was discarded in favour of a new and more accurate clock. In 1928, the remains of the original clock were found in the clock tower of the cathedral and finally, in 1956, the clock was restored to close to its original condition, which involved reproducing parts of the timing mechanism that had been lost.

The early timing mechanism involved a combination of two components known as a *verge escapement* and *foliot*. Figure 3.2a shows the verge escapement mechanism. The aptly named *crown wheel* is a toothed wheel with an odd number of teeth, where the teeth are arranged in a ring like the ornamentation of a regal crown. The teeth are shaped with one steep side and the other side sloping. The *verge* is a rod to which two plates called *pallets* are attached that are inclined to each other as shown; in the figure, pallet A is situated in one of the

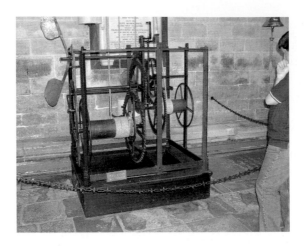

Figure 3.1   The reconstructed Salisbury Cathedral clock.

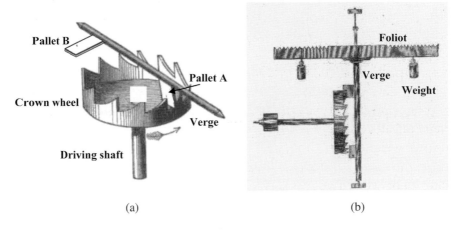

(a)	(b)

Figure 3.2   (a) The verge escapement. (b) The verge escapement and foliot.

gaps between the teeth. The drive shaft is rotated by the fall of a heavy weight attached to a rope wrapped round the shaft and as it rotates it pushes pallet A forward causing it to climb up the sloping face of the tooth ahead and causing the verge to rotate. As it rotates, pallet B will move to fall into a gap on the other side of the crown wheel, while pallet A completely disengages. Now the action is repeated with the pallets changing roles but the net effect is that the verge continuously rocks to and fro. If that were the complete arrangement, the rate of the rocking motion would depend on the weight providing the motion, the diameter of the drive shaft around which the rope is wrapped and the friction between the pallets and the edges of the teeth. It is the inertia of the foliot, two masses attached to a heavy arm (Figure 3.2b) which controls the rocking rate. If the foliot were absent the drive shaft would rotate very quickly with the crown wheel and pallet making a rattling noise like a machine gun. At the other extreme, with a very massive foliot the rocking motion would be very slow and the drive shaft would rotate slowly. The inertia of the foliot can be altered to adjust the rate of the clock by moving the weights in and out on the foliot arm.

The combination of verge escapement and foliot causes the drive shaft to rotate in a spasmodic fashion but at a steady rate and this motion, acting through a series of gear chains, controls the mechanism indicating the time. The Salisbury clock did not have a clock face, so that the time could not be seen, but was designed to strike the hours, with the number of strikes indicating the time — sufficient to inform the clergy when to assemble for their daily prayers. In this sense it was a true *clock*, a name derived from the French *cloche* meaning 'bell'. Timing devices that do not chime are technically *timepieces* but it is common practice to use the word 'clock' for any non-portable device that indicates the time. In Figure 3.1, the left-hand side of the clock with its own weight-and-rope system controlled the striking mechanism and the right-hand side the clock mechanism. Every 12 hours the weights have to be wound up to maintain the clock in continuous operation.

At best, a clock such as that at Salisbury could be accurate to about two minutes per day, although an accuracy of about 15 minutes a day would be more common. The performance would vary through the year with the prevailing temperature, which would cause the metal components, particularly the foliot, to expand and contract.

3.2 Pendulum-Controlled Clocks

In 1582, the Italian mathematician and astronomer Galileo Galilei (1564–1642) observed the swinging motion of a chandelier in Pisa cathedral. He noticed that the period of its swing, which it is said he timed using his pulse, did not change as the amplitude of the swing died down. In 1602, he carried out a series of experiments on the properties of the pendulum and later, in 1637, he conceived the idea of using a pendulum to regulate a mechanical clock. His illegitimate son, Vincenzo Gamba (1606–1649), started to construct such a clock but died before the work was completed.

Galileo Galilei   Christiaan Huygens

A few years after this attempt, in 1656, a pendulum clock was invented by the Dutch scientist Christiaan Huygens (1629–1685) and constructed under his direction by the Dutch clockmaker Salomon Coster (1620–1659). This clock used a verge escapement with the pendulum in the form of a rigid rod with a weight attached to the end controlling the rate of rotation of the driving shaft, much as the foliot had done previously (Figure 3.3). The pendulum was kept swinging by the energy provided by the falling weight through the verge, just as the foliot maintained its rocking motion.

From Figure 3.2a, it will be seen that the angle between the planes of the two pallets is quite large, in the order of 120°, and for the verge escapement mechanism to operate the pendulum would have to swing through a similar angle. Although Galileo detected no change in the period of the cathedral chandelier as its amplitude died down, there was a change, although not one he could detect by crudely measuring time using his pulse. The period of a pendulum varies with the angle

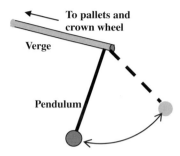

Figure 3.3 A pendulum attached to a verge escapement.

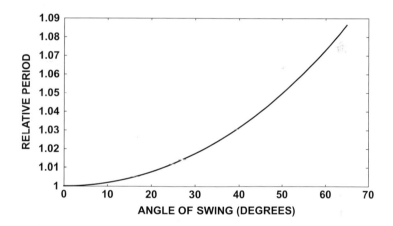

Figure 3.4 The variation of the period of a simple pendulum with the amplitude of the swing.

of swing, θ, away from the vertical as shown in Figure 3.4. For very small-amplitude[1] swings, the period varies little with angle but it varies much more rapidly with angle for large swings, the angle for a verge escapement being about 60°. There are small variations in the amplitude of the period swing due to slight differences around the crown wheel and the corresponding variation in period, although small, is significant for a clock designed to operate over long periods of time.

---

[1]Amplitude is the maximum deviation from the mean position, i.e. 60° for a total swing through 120°.

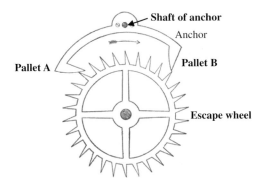

Figure 3.5   An anchor escapement.

Another problem with having a wide amplitude swing is that a small-scale clock, for use in the home, would have to be made very wide to accommodate the lateral motion of the pendulum. These problems led to improved designs of escapements that only required small amplitudes of swing. The first of these was the *anchor escapement* (Figure 3.5), probably invented by the eminent British scientist Robert Hooke (1635–1703). The earliest known clock with an anchor escapement, constructed in 1670, is that installed in the clock tower of Wadham College, Oxford.

In Figure 3.5, pallet A is just disengaging from a gap in the toothed *escape wheel* and as it does so, pallet B is moving downwards to become engaged. When the pendulum swings back the other way, pallet B will disengage and pallet A will re-engage with the escape wheel. The pendulum is connected to the shaft of the anchor so that as it swings to and fro, it rocks the anchor. In its turn, the motion of the anchor is linked by a mechanical arrangement that transmits energy to the pendulum by giving the pendulum a periodic impulse to keep it swinging.

A pendulum with an anchor escapement needs to swing with an amplitude of only 5° and this gives much better timekeeping if there are small variations in amplitude for any reason. Again, the small swing means that the pendulum could be longer within a narrow clock and around 1680 the British clockmaker William Clement (1638–1709) designed and built the first *longcase clock* or, as they came to be called, *grandfather clock*. The pendulum in a grandfather clock, of one metre length, has a period of two seconds, which means that if the escape wheel has 30 teeth, then its shaft rotates once per minute and can be used through simple gearing to drive the minute hand of the clock. The anchor escapement made clocks sufficiently accurate to justify the use of a minute hand; before about 1690, it was exceptional to find a clock with anything other than an hour hand.

A serious problem with the anchor escapement is that during each cycle there is a slight reversal of the motion of the escape wheel, called *recoil*, which causes wear on its teeth. For this reason, a modified form of escapement called the *deadbeat escapement* is preferred because it does not have this problem. The improvement is due to a change in the shape of the pallets, which prevents the reverse motion, but the essential action is the same as that of the anchor escapement.

A final problem associated with pendulum clocks is that of the change in the length of the pendulum due to expansion and contraction with changes of temperature. This problem has been solved in two ways that use the differential expansion of different materials. The first way, illustrated in Figure 3.6a, is the *mercury pendulum*, invented in 1721 by the British clockmaker George Graham (1673–1751). The weight at

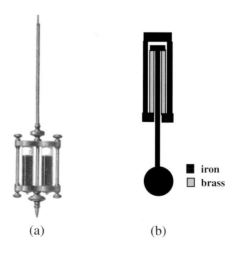

(a)   (b)

iron
brass

Figure 3.6   Temperature-corrected pendulums. (a) The mercury pendulum. (b) The gridiron pendulum.

the end of the pendulum rod consists mainly of tubes of mercury. The expansion of the main shaft of the pendulum lowers its centre of mass but this is compensated for by expansion of the mercury, the centre of mass of which rises when it expands.

The second way is to use the *gridiron pendulum* (Figure 3.6b), invented in 1726 by the eminent British clockmaker John Harrison — to whom George Graham was friend, supporter and mentor — whose further achievements will be described in Sections 3.3.1 and 3.3.2. In this arrangement, the downward movement of the centre of mass of the pendulum due to expansion of the iron bars is compensated for by its upward movement due to expansion of the brass bars.

The very best pendulum clocks are capable of great accuracy, with an error approaching one second per year, and they were the standard timekeeping devices used worldwide until the invention of the quartz clock in 1927 (Section 4.2).

3.3 Navigation and Time

It was October 22nd 1707 and the fog that enveloped the whole region of the western approaches to the English Channel made visual navigation impossible. The 16 ships of Admiral Sir Cloudesley Shovell's flotilla were making their way home after a victorious campaign in the Mediterranean, part of the War of the Spanish Succession (1700–1714); he would surely receive the hero's welcome that he so richly deserved. But that was not to be. The ships were hopelessly off course and instead of being just off the coast of France, as they imagined, they were much further to the west. They set their course to the north-east, confident that they were entering the mouth of the Channel, led by Sir Cloudesley's own ship, the *Association*, which was the first to strike the rocks off the coast of the Scilly Isles. Three more ships shared the fate of the *Association* and nearly 2,000 men perished in that terrible disaster.

The disaster came about because of the inability of those navigating the sea to know exactly where they were. Given a sight of the Sun, they could determine their latitude with some precision. The latitude of the Sun's orbit at any time of the year was known and by the use of a sextant

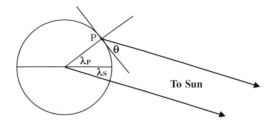

Figure 3.7 Determining latitude by observing the altitude of the Sun at its zenith position.

the altitude of the Sun when it was at its zenith could be determined. This is shown in Figure 3.7, where the Sun is (apparently) orbiting the Earth at latitude λ_S and at the local noon, when it is at its zenith, its altitude above the horizon is θ. It is straightforward to determine the latitude of point P as

$$\lambda_P = \pi/2 - \theta - \lambda_S. \tag{3.1}$$

If the altitude of the Sun could be determined with an accuracy of one minute of arc then the latitude would be known within 2 km, which would serve for most purposes up to the 18th century.

The real problem was to determine the longitude for which observations of the Sun alone were ineffective. What can be determined is when the local noon occurs, because then the Sun is at its maximum altitude. Since the Earth spins on its axis at a steady rate, the longitude would be given by the time difference between noon at Greenwich, from which longitude is measured, and noon where the solar observation was made. For example, if it was known that your local noon occurred when the time at Greenwich was 6 pm, then you were at a longitude of $(6/24) \times 360° = 90°$ W. This shows that all you have to do to know your longitude is to know the time in Greenwich when you observe your local noon. The solution is simple; set a clock to Greenwich time and take it with you when you go on your voyage. The difficulty with that solution is that a pendulum clock under the turbulent conditions that often prevail at sea is an extremely unreliable and erratic keeper of time; an error of 1 minute at latitude 45° would entail an error of 20 km in estimated position, but errors could be very much greater than that.

The problem was well known before Admiral Shovell's disaster and many eminent men had given it their attention. Galileo had suggested that the orbits of Jupiter's satellites could provide the solution. They orbit with great regularity and the times that they disappeared behind, and reappeared from, Jupiter's disk gave an absolute measure of time anywhere on Earth. Actually, because of the motions of both the Earth and Jupiter, sometimes these bodies approach each other and at other times they become further apart. Because of this variation in their distance and the finite speed of light, there can be a drift of several minutes between the appearance and disappearance of the satellites. Given that observing these satellites was not a simple matter in any case, this method was never seriously considered. Other eminent people from time to time sought other solutions that depended on observing heavenly bodies that moved on predictable paths. One solution, suggested to King Charles II by a Frenchman, Le Sieur de St-Pierre (1701–1755), involved observing the position of the Moon against the background of fixed stars. The king was impressed by this idea but it required that the positions of stars be known accurately, which they were not at that time. John Flamsteed (1646–1719), the first Astronomer Royal, took advantage of this opportunity and, shortly after his appointment in 1675, he persuaded the king to set up the Royal Greenwich Observatory for the purpose of accurately determining the positions of fixed stars. Eventually, he produced a catalogue of the positions of 3,000 stars.

An astronomical method that came to be widely used to determine longitude is the *Method of Lunar Distances*. This consisted of measuring the angle between the Moon and some other celestial object, which could be a star or the Sun. As an example, the combined motion of the Earth around the Sun and of the Moon around the Earth results in the Moon moving through one complete cycle, 360°, relative to the Sun in about 28 days or about 0.5 minutes of arc per minute of time. This is illustrated In Figure 3.8. Starting at positions 1, when the Moon and Sun are in the same direction, after about nine days positions 2 of the two bodies give an angle θ between them. After 28 days they are collinear again.

If the angle between the Sun and the Moon can be measured to an accuracy of 0.5 minutes of arc it gives the time to within one

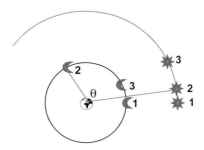

Figure 3.8 The Method of Lunar Distances. For clarity, the lunar and solar orbits are not to scale.

minute which, as indicated previously, corresponds to an uncertainty of about 20 kilometres at mid-latitudes. However, since the orbits of the Moon round the Earth and of the Earth round the Sun are not circular, complicated corrections have to be made to get a reasonable estimate of time.

As a consequence of the loss of many ships due to navigational errors, in 1714 the British parliament passed the Longitude Act which set up a Board of Longitude, a panel of scientists, naval men and government officials, whose job it was to judge proposals for determining longitude. For anyone who could design a foolproof scheme to determine latitude to within half a degree (nearly 40 km at mid-latitudes) the prize was £20,000, an enormous fortune equivalent to nearly £3,000,000 today. We can judge the poor state of navigation in those days by the large distance error that would qualify to win the prize. There were lesser prizes for less accurate solutions — £15,000 for an error less than two-thirds of a degree and £10,000 for an error less than a degree. The British government, with the largest merchant fleet and largest navy in the world, was desperate to solve this problem.

The solution to the longitude problem within the limits set by the Board of Longitude was eventually solved by the most straightforward approach, the design of a timepiece that would keep reasonably accurate time at sea.

3.3.1 *John Harrison and his sea clocks*

John Harrison (1693–1776), from Foulby in Yorkshire, was a carpenter who developed an interest in making clocks. In 1713, he constructed a longcase clock using only the material he best understood at that time — wood. The woods he used were oak and *lignum vitae*, an extremely hard wood of West Indian origin that had the advantage that it exuded sap and was self-lubricating. The lubrication of clocks was a tricky operation in those days and the oils used were as likely to clog up the mechanism as to aid its motion and reduce wear. Three of Harrison's early wooden clocks still survive; one is in the care of the Worshipful Company of Clockmakers, another in the Science Museum in London and the third in Nostell Priory in Yorkshire. Later, together with his brother James, he made other clocks, also still surviving in various locations, with wooden movements but incorporating the gridiron pendulum. John was a man who was never satisfied with what he had achieved, or what already existed, in the way of technical devices and he always sought to do better. He improved on the anchor escapement, with its inevitable wear, and invented the *grasshopper escapement*, a complex design that was virtually friction free.

John Harrison

Harrison decided to take up the challenge presented by the Board of Longitude by building a clock that would be accurate and operate reliably under the turbulent conditions encountered at sea. Clearly, a pendulum clock would not work; a pendulum operates under the influence of gravity and on a rolling and pitching ship the direction of gravity relative to the pendulum constantly changes. His first *sea clock*, now known as H1, was essentially a portable version of his precision

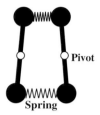

Figure 3.9   A schematic representation of the dumbbell oscillator in H1. The actual device was far more elaborate in appearance.

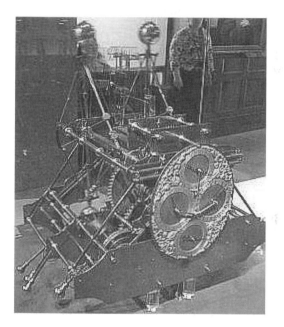

Figure 3.10   Harrison's H1 sea clock.

wooden clocks. Instead of using a pendulum to give a standard time interval, he used two dumbbell-like devices linked by springs, as shown in Figure 3.9. The dumbbells oscillated about the pivots, propelled by the forces due to the alternating compressed and stretched springs. The device would work in any orientation and was completely independent of the strength or direction of the gravitational force.

The clock weighs about 35 kg and is about 63 cm tall. Figure 3.10 shows the clock with weights of the dumbbell oscillator clearly visible at the top of the image. The Board of Longitude was sufficiently impressed by H1 to subject it to a sea trial and in 1736, it was sent on a round trip to Lisbon and back. The test was satisfactory but the definitive test demanded by the Board was a return journey across the Atlantic. Anyway, the Board thought that H1 was promising and gave Harrison a grant of £500 to develop his ideas further. Five years later, Harrison had constructed H2, a more robust version of H1, but it could not be tested because Spain and Britain were at war (War of the Austrian Succession) and the Board did not want to run the risk of this important development falling into the hands of the enemy.

At about this time, Harrison spotted a flaw in his design in that the dumbbell timing device would be affected by the pitching motion of a ship. So, with another grant of £500 from the Board, in 1741 he set about constructing H3, a task that would take him 17 years. There were several changes to the design of H2. Instead of linked dumbbells, it had two *balance wheels* (Section 3.3.2) linked by a spiral spring of 1½ turns. He also designed enclosed roller bearings (the forerunner of modern roller bearings and ball bearings) to reduce the friction in the oscillation of the balance wheels and used a bimetallic strip to correct for the effect of temperature changes on the spiral spring. A bimetallic strip, illustrated in Figure 3.11, consists of two rigidly connected strips of different metals, e.g. steel and brass, with different coefficients of expansion. If metal A has a higher coefficient of expansion than metal B, then when the temperature increases the combination will bend as shown. With one end fixed, the motion of the other end can be used to control the device in which it is installed. It is a common component in many modern appliances, e.g. electric toasters.

After all his labours in making H3, Harrison came to the realization that a sea clock was not the answer to the problem. A successor to his mentor George Graham, Thomas Mudge (1715–1794), had made

Figure 3.11   The action of a bimetallic strip. When heated the strip bends.

a watch that was just as good a timekeeper as Harrison's massive sea clocks. The characteristic that made the watch so successful is that the balance wheel oscillated much faster than the timing-control mechanisms in the clocks, and this reduced the effect of motion while it was working. Harrison then decided to change tack completely and apply his ingenuity to making a watch better than any previously made.

3.3.2 *Harrison and his watches*

Harrison took six years to construct his first *sea watch*, generally known as H4 (Figure 3.12). It resembles a normal pocket watch but is much larger, about 13 cm in diameter. As in all pocket watches, the timing device was a *balance wheel*. Figure 3.13 shows a simple balance wheel

Figure 3.12 Harrison's sea watch, H4.

Figure 3.13 A simple balance wheel.

from a modern inexpensive clock which illustrates the essential features of the device. The shaft of the wheel engages with jewelled bearings that reduce frictional forces. The *hairspring* provides the periodic force that rotates the balance wheel with an amplitude of about 270° in each direction from the neutral position. Temperature affects the period in three ways, firstly by lengthening the spring, secondly by varying the elasticity of the spring and thirdly by changing the dimensions of the balance wheel. Modern watches obviate these problems by the use of materials that have zero coefficient of expansion for both the balance wheel and the spring and also with unchanging elasticity of the spring material. Harrison's solution was the bimetallic strip. At one end, the hairspring had a straight portion and as the temperature varied, the bimetallic strip pressing against this section varied the period to provide what is called a *compensation curb* that compensated for the variation in period due to the other temperature effects. The escapement mechanism for H4 was a variation of the verge escapement that allowed for the very wide swings of the balance wheels and incorporated pallets made of diamond that made them free of frictional wear.

The driving force of the watch was a wound-up spring, which introduced the problem that as the spring ran down, the force it applied varied and this slightly affected the period of the balance wheels. To compensate for this, the watch had a *remontoire*, which is a small secondary spring that powers the escapement and is rewound at frequent intervals by the mainspring. The force applied by the remontoire is much more constant than that from the mainspring itself.

In 1761, the watch was completed and was sent by ship, under the care of Harrison's son William, to Jamaica for a trial. It was just five seconds slow when it reached Jamaica, corresponding to a longitude error of about 1.6 kilometres. Harrison expected to receive the prize offered by the Board of Longitude, but they declined to award it and insisted on another trial. One member of the Board was Nevil Maskelyne (1732–1811), a strong proponent of the Method of Lunar Distances, who persuaded the Board that the result was a fluke that had come about by fortuitous cancelling out of large errors of timing. A second sea trial, this time to Barbados, gave a longitude error of

16 kilometres, still within the bounds set by the Board, and in 1765 the Board decided that Harrison should receive the prize — £10,000 immediately, with the remaining money to be handed over when the design was made available to other watchmakers so that it could be reproduced. However, this offer was conditional on approval by the Astronomer Royal, who would further test the watch. By this time, Maskelyne had been made Astronomer Royal and once again claimed that the results of the trials were inconclusive, so H4 was never adopted for general use.

While all this was going on, Harrison was working on another watch — H5. By this time, Harrison was thoroughly disillusioned about the objectivity of the Board and sought, and was granted, an audience with King George III to press his case. The king was supportive and tested H5 himself, finding that over a six-month period it kept time to within one-third of a second per day. Eventually, in 1773, due to the king's influence, Harrison received £8,750, but he never received the full prize that he had clearly earned.

The precise watches, eventually called *chronometers*, were initially so expensive that they were not widely used. For many years, the Method of Lunar Distances was used in parallel with chronometers until, in the early part of the 19th century, the price of the chronometers fell to a more affordable level, in the order of £50 — although still equivalent to about £10,000 today.

3.4 Modern Mechanical Watches

It is surprising that the vast majority of modern watches, certified as chronometers, do not match the performance that Harrison's watches were claimed to have reached. A watch is given chronometer status by the COSC (Contrôle Officiel Suisse des Chronomètres) if it maintains an accuracy of between −4 and +6 seconds per day, a standard that would not have won the prize offered by the Board of Longitude. This modern definition of a chronometer puts John Harrison's achievements into perspective. Of course, the very best mechanical chronometers made today can match those of John Harrison, but there are now new ways of recording time that no mechanical system can ever match.

Chapter 4

Modern Timekeeping

4.1 Standards of Time

The need for agreed international standards for various basic units was realized during the 18th century when the physical sciences were developing rapidly and were becoming quantitatively precise. In the late 18th century in revolutionary France, units of length and mass were established. The metre was defined as one-ten millionth of the distance from the equator to the North Pole and the kilogram was taken as the mass of one litre of water, where one litre is one-thousandth of a cubic metre. The standards were established in material form by a platinum bar of length one metre and an ingot of a platinum-iridium alloy of mass one kilogram placed in the Bureau International des Poids et Mesures (International Bureau of Weights and Measures) in Sévres in France. Replicas of these standard units were then distributed from this source. The basis of these standards is that they were meant to be dependent on quantities that were immutable and agreed by everyone — the distance from the equator to the North Pole, for example. However, although that distance has been refined with time, the standard as established in Sévres remained unchanged. The mass of an object is an invariable quantity, but its length can vary with physical conditions so the modern standard of length is based on the speed of light which, in a vacuum, is absolutely constant. The standard metre is now defined as $1/299,792,458$ of the distance that light travels in one second — but that definition depends on the establishment of a standard for the second, something we now consider.

The unit for time was much more challenging to establish. Up to about the end of the 17th century it was the second, derived from the *solar day*, with 24 hours divided into minutes and seconds, which was the time between successive noontimes, when the Sun was at its zenith. Because of the Earth's elliptical orbit around the Sun, the solar day varied by several seconds throughout the year, but when measurements could only be made with very limited accuracy that somewhat variable definition was good enough. The 24 hours depended on the rotational speed of the Earth, which was taken as constant, with the variation in the length of a day being due to the Earth's elliptical orbit. Later, a new standard second was established based on *sidereal time*, which was the period between seeing a star at the same position in the sky on successive nights. This period, 23 hours 56 minutes 4.09 seconds, was dependent only on the uniform rotation of the Earth and also had the advantage that observing a star, essentially a point source of light, could be made much more accurately than observations of a visually large object like the Sun.

It became clear at the beginning of the 20th century that the rotation of the Earth was not uniform. Due to various factors, in particular convection currents in the Earth's interior that changed the distribution of matter within it, the length of a day could vary by a tiny amount away from the value predicted by taking into account the Earth's eccentric orbit. In 1952, the International Astronomical Union adopted a new standard of time, *ephemeris time*, based on the orbital period of the Earth for the year 1900 so that the standard second became 1/31,556,925.9747 of that year. That is not the end of the story for establishing a standard second — something that will be described in Section 4.3.

4.2 Quartz Clocks

It was the advent of the quartz clock that introduced a new factor into the assessment of the accuracy of timekeeping, which was that the rate at which the clocks ran was more uniform than the rate at which the Earth spun on its axis. These changes in spin rate were unpredictable and posed the problem of establishing a standard against which to assess the accuracy of quartz clocks, giving rise to the definition of ephemeris time. We shall return to the question of assessing the performance of quartz clocks after describing the physical principles on which their operation is based.

4.2.1 *The piezoelectric effect*

Quartz is a common naturally occurring mineral (Figure 4.1) with chemical composition SiO_2, two atoms of oxygen associated with a silicon atom, which has the chemical name *silica*. The fundamental unit of the quartz structure actually consists of a tetrahedral arrangement of oxygen atoms attached to silicon, the SiO_4 unit (Figure 4.2a), but these

Figure 4.1   Naturally occurring quartz crystals.

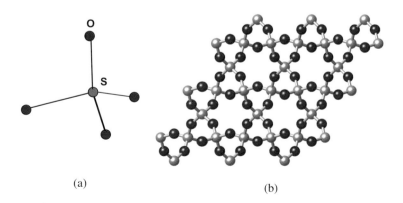

(a)

(b)

Figure 4.2   (a) The SiO_4 unit. (b) The quartz framework structure.

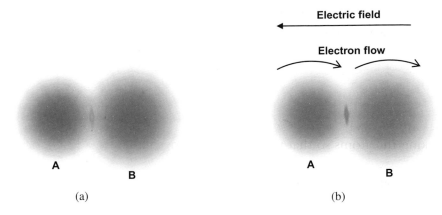

Figure 4.3   (a) The electron cloud associated with a pair of bonded atoms. (b) The change of the electron density between the atoms due to an imposed electric field.

are linked together in a three-dimensional framework in such a way that each oxygen atom is part of two tetrahedral units (Figure 4.2b), thus leading to the overall SiO_2 composition. Pure quartz is colourless, but with traces of impurity it appears as coloured semi-precious stones, such as *citrine*, *amethyst* and *rose quartz*, which are used for jewellery.

The property of quartz that enables it to be used for making accurate timepieces is that it is a *piezoelectric* crystal. Piezoelectric materials have the property that when they are placed in an electric field they undergo a mechanical strain and, conversely, when the material is mechanically strained it becomes electrically polarized and generates an electric field in its vicinity. Let us see what mechanism can produce this effect. In Figure 4.3a there is a representation of a pair of dissimilar atoms with the associated electrons forming a fuzzy cloud around each atomic nucleus. The two atoms are bonded together so that the electron density is different from the sum of that for two isolated atoms and there is a slight build-up of electron density between the two nuclei. The attraction of the positively-charged nuclei towards the concentration of electron density gives the bonding. If the pair of atoms is now placed in an electric field, then this will lead to a displacement of the negatively-charged electron cloud with respect to the nuclei. Since the direction of an electric field is the way a positive charge would move, the direction of migration of the negatively-charged electron cloud is as shown by the arrows in Figure 4.3b. It may happen that the electrons of atom B are more strongly bound to the nucleus and will be less easily displaced than those of atom A. The consequence of this is that less of B's electron cloud will move out of the space between the nuclei than will move in from A, so that the concentration of density between the nuclei will be greater. This will strengthen the bond linking the atoms so that the nuclei will move closer together. Within a complex material there will be many bonds at various orientations with respect to the field, some of which will become longer and some shorter. For some crystals with a particular symmetry for every pair of atoms AB there is another parallel pair BA so that the stretching of one bond is equal in magnitude to the shrinking of the other. In that case, the material is not piezoelectric and when placed in an electric field, there is no external change in the dimensions of the material. However, for quartz as for many other materials no such balancing of stretching and shrinking occurs and the dimensions of a quartz crystal do change in an electric field — i.e. it experiences a mechanical strain.

The piezoelectric effect is reversible, giving the *inverse piezoelectric effect*, so that if the material is physically strained, the electric charge will flow in and out of the region between the nuclei. The net flow of charge will lead to a polarization of the material and hence a field will be produced in its vicinity.

4.2.2 *Resonance*

The phenomenon of resonance is one that is common in everyday life and may be illustrated by the behaviour of a child's swing. If the swing is set into motion and left to its own devices the amplitude of the swing will gradually reduce due to air resistance and possibly friction, depending on how the swing is constructed. However, if the swing is given a small push forward whenever it reaches the end of its backward motion then the energy provided will compensate for the loss of energy and the swing will operate continuously at constant amplitude.

In the case of the swing care must be taken not to push too hard, otherwise the swing's amplitude of motion may become too great, which would be hazardous for the child and perhaps the structure of the swing itself. There are well-known examples of destructive resonance. If a wine glass is tapped with a spoon it will emit a musical note of a particular frequency. It was said that the opera singer Enrico Caruso (1873–1921) could shatter a wine glass by singing a constant note in its vicinity at its natural frequency. Another well-authenticated example occurred in 1831, when a party of 74 soldiers crossed the Broughton Suspension Bridge that crossed the River Irwell in Salford. As they crossed, the bridge began to rock in resonance with their regular marching pace and just before they reached the end of the bridge one of the suspension chains snapped and the bridge collapsed, fortunately without any loss of life, although there were a number of injuries.

Piezoelectric crystals are used to produce ultrasound devices that are used for medical diagnosis. A typical device is shown schematically in Figure 4.4. The piezoelectric material, in the form of a plate, is sandwiched between two electrodes connected to a source that provides an alternating field with a frequency that resonates with the natural frequency of vibration of the plate. The plate vibrates and emits ultrasonic sound waves. The backing plate is there to absorb the

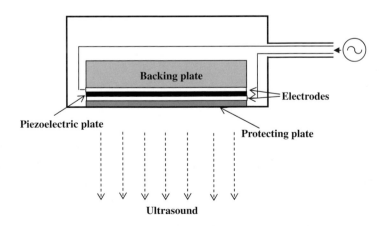

Figure 4.4   A piezoelectric ultrasound generator.

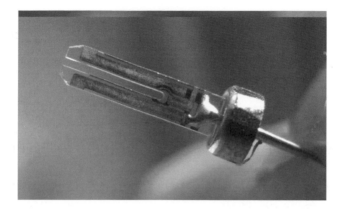

Figure 4.5   A quartz crystal in the shape of a tuning fork.

ultrasound going in that direction so it is not reflected back to interfere with the ultrasonic waves coming from the front surface.

4.2.3 *The essential mechanism of a quartz clock or watch*

At the heart of a quartz timepiece is a piece of quartz so shaped that it has some convenient natural frequency, usually chosen as 32,768 Hz[1] (2^{15} Hz) because the vibrations are counted by digital counters that operate best as binary devices, i.e. in powers of 2. The shape is often in the form of a tuning fork some three millimetres long, as shown in Figure 4.5.

We have seen from the ultrasound generator that a piezoelectric crystal subjected to an alternating field with the same frequency as the natural frequency of the crystal will be set into vibration at its natural frequency. The inverse piezoelectric effect indicates that a piezoelectric crystal vibrating at its natural frequency will generate an alternating field with a frequency equal to that of the vibrating crystal. In essence, the action of a quartz oscillator in a quartz clock is to take a current from the alternating field of the vibrating crystal, then to send this current into an amplifier powered by a battery, with the amplified current providing

---

[1] One hertz (Hz) is a frequency of one cycle per second, e.g. 50 Hz is 50 cycles per second, the electric mains frequency in the UK.

an alternating field across the crystal to maintain its vibrations. It is an example of what is called *positive feedback*, where the amplified output of a device is fed back into the device to generate further output. It is this kind of positive feedback that gives an oscillating circuit; in general, it is the combination of electrical components in the circuit — inductors and capacitors — that give the circuit its natural frequency, but in this case it is the quartz crystal.

Sound vibrations from a resonating circuit driven at 32,768 Hz are beyond the normal audible range, so do not create a noise nuisance, but are of low enough frequency to be recorded by fairly simple and therefore cheap digital counters. After 2^{15} cycles the counter will overflow and this will trigger a one second pulse that can be used to control either a digital or an analogue display of time — the digital usually as a liquid crystal display, which consumes little power, and the analogue as the motions of second, minute and hour hands.

4.2.4 *The accuracy of a quartz timepiece*

Quartz is an ideal material for use as a frequency control for timepieces. It is robust and has a low coefficient of thermal expansion, which means that the natural frequency of a shaped quartz crystal will be little affected by modest variations of temperature. However, for the very highest accuracy as required by institutions which act as centres for the establishment of time standards, quartz clocks were kept in constant-temperature enclosures. We say *were kept* because, as we shall see, they are no longer the means of establishing an international time standard.

The machining of a quartz crystal to achieve precisely the natural frequency of 32,768 Hz can be difficult and expensive, so for cheaper timepieces the technique of *inhibition compensation* is used. The crystal is machined to give greater than 32,768 Hz and then is tested on equipment that compares it with a reasonably accurate time standard. The same equipment automatically provides the timepiece with a logic circuit that every so often, perhaps every ten minutes, skips a few cycles to give the right average of 32,768 Hz. This means that the timepiece does not run at a completely uniform rate, but for most day-to-day use this is not important. Although, as described, the circuitry may seem to be complicated, it is normally all achieved with printed circuits and fits comfortably within the confines of a wristwatch.

For a normal medium-quality wristwatch, the accuracy is in the order of ±½ second per day or about 3 minutes per year. We have already mentioned the difficulty of defining the accuracy of quartz clocks. Although ephemeris time defines a standard second, it does not provide a piece of equipment against which to measure the accuracy of a timepiece. However, testing high-quality quartz clocks kept in constant temperature enclosures by comparing them with each other suggests that their accuracy will be in the order of 2 milliseconds per day, or less than one second per year.

Quartz clocks and wristwatches now dominate the market for domestic and commercial use. They are inexpensive and sufficiently precise for almost all normal requirements. However, there are requirements for which they are inadequate and we now describe the ultimate timepiece for accuracy, the atomic clock.

4.3 The Atomic Clock

From the first use of the verge escapement and foliot, the basis of timekeeping has depended on recording the number of oscillations of some periodic motion — a foliot, pendulum, balance wheel or the vibration of a quartz crystal. All these methods of measuring time are based on some production process that must introduce an element of variability in their performance and they are affected by prevailing conditions, such as temperature or air resistance. Although production processes can be made as precise as possible and steps can be taken to mitigate the effects of variations in the prevailing conditions, the performance of the resultant timepiece must have some limitation of accuracy. With a first-class quartz clock time can be kept with an accuracy of less than a second per year, which might be thought as sufficient for all human activities. However, for modern requirements, such as the use of Global Positioning Systems (GPS), which rely on knowing the positions of Earth satellites moving at eight kilometres per second, with both civilian and military applications, an error in timing as small as a millisecond — during which time the satellite moves by eight metres — is unacceptable. This kind of need is met by an atomic clock, the physical basis of which will now be described.

4.3.1 *Radiation from atoms and molecules*

The electrons in atoms can only have specific energies, a result that comes from the requirements of quantum mechanics, which describes the behaviour of atomic and nuclear systems. When an electron moves from one of its allowed energy levels with energy E_1 to a lower energy level with energy E_2 the energy released appears as a packet of electromagnetic radiation, referred to as a *photon*, the frequency v, of which is given by

$$hv = E_1 - E_2, \qquad (4.1)$$

where h is *Planck's constant*, 6.626×10^{-34} J s.[2] When the difference in energy is sufficiently large, the radiation produced will be in the ultraviolet region, while for a smaller difference it can be visible light, giving the characteristic visible spectra of atoms. However, what are of interest for the construction of an atomic clock are smaller differences in energy, some 10^5 to 10^6 times smaller, giving frequencies in the microwave range. These come about from transitions between electron states differing slightly in their energies due to different interactions of an electron with the internally generated magnetic fields within an atom which are due to the nucleus and all the electrons. Such small energy differences give what is known as *hyperfine structure* and this can also exist in the energy states of molecules, consisting of two or more atoms, due to different ways in which the magnetic properties of different nuclei interact. The microwave range of frequencies is roughly between 300 MHz and 300 GHz.[3] For atomic clocks, one commonly used radiation is that of a hyperfine transition in a caesium isotope, Cs-133, with a frequency of 9,192,631,770 Hz (9.192631770 GHz), although other radiation is sometimes used.

We have already commented on the fact that the use of foliots, pendulums, balance wheels and machined quartz crystals to provide

---

[2] The joule (J) is the Standard International (SI) unit of energy.
[3] 1 MHz (megahertz) is 10^6 Hz and 1 GHz (gigahertz) is 10^9 Hz.

some standard period for the recording of time has the drawback that these are man-made objects and, no matter how much skill is employed in their construction, no two will ever be *precisely* the same. The frequencies of radiation from the energy transitions of electrons in an atom are identical for all samples of an isotope of the same kind; there cannot be the slightest variation. In addition, the frequencies of these transitions are not affected by prevailing conditions, such as temperature, and so provide a very precise way of defining and measuring time. The current internationally recognized standard second is defined as the time for 9,192,631,770 cycles of the radiation from the previously mentioned hyperfine transition of Cs-133. Of course, to fix this standard one originally had to refer to the agreed astronomical time scale, ephemeris time, but having done that, the standard is now the purely physical one based on the Cs-133 transition.

4.3.2 *The mechanism of an atomic clock*

The fact that the radiation from atoms could provide an absolute standard of frequency and hence a precise way of measuring time was originally suggested by the British physicist William Thomson (Lord Kelvin; 1824–1907), but the technology of the time could not use the phenomenon. The first atomic clock was built in 1949 by the US National Bureau of Standards (now National Institute of Standards and Technology) and was based on microwave radiation from the molecule of ammonia, NH_3. It was not as accurate as the existing quartz clocks but it pointed the way towards constructing something better. There are several designs of atomic clock; the most recent and accurate is the *caesium fountain clock*, examples of which have been constructed at the national standards institutes of the US, the UK, France and Germany. The essential design of a caesium fountain clock is shown in Figure 4.6; we shall describe its mode of operation step by step, explaining the essential physics as we go along.

We have previously mentioned that the frequency of the radiation *emitted* by an atom in moving from one particular energy state to a particular lower one is unaffected by temperature, but temperature will affect the frequency *seen* by an observer. Temperature is a measure of

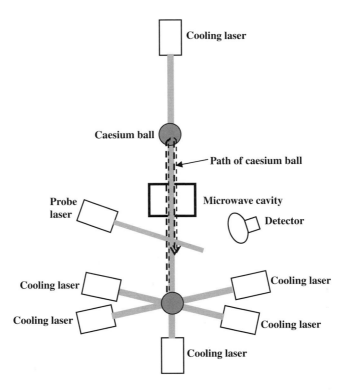

Figure 4.6   A schematic of a caesium fountain atomic clock.

the kinetic energy (energy of motion) of the material, the individual atoms or molecules of which are in random motion at various speeds in all directions. The higher the temperature, the faster the atoms move. For coherent particles (atoms or molecules) of mass m at temperature T the average (actually root-mean-square) speed of the atoms moving in random directions is

$$v_{rms} = \sqrt{\frac{3kT}{m}}, \tag{4.2}$$

in which k is the Boltzmann constant, 1.381×10^{-23} J K^{-1}.[4]

   If a source of electromagnetic radiation, or indeed of any wave motion, such as a sound wave, moves away or towards an observer,

---

[4] The kelvin (K) is the unit of absolute temperature, which is proportional to the energy of the particles of the material at that temperature.

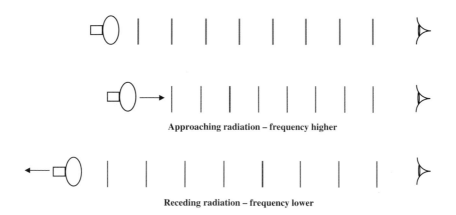

Approaching radiation – frequency higher

Receding radiation – frequency lower

Figure 4.7 A schematic representation of the Doppler effect.

then the frequency detected is different from that being emitted due to a phenomenon known as the *Doppler effect*, illustrated schematically in Figure 4.7. When the radiation source is approaching, it shortens the time between the arrival of successive peaks and the frequency detected, and hence the effective energy of the photons is higher. Conversely, when the radiation source is receding, it lengthens the time between the arrival of successive peaks and the frequency detected, and hence the effective energy of the photons is less.

Due to the Doppler effect, when radiation from a large number of atoms in random motion is received, a range of frequencies around the true frequency is detected — in technical language the spectral line is broadened. For an atomic clock to be accurate, it is essential that the spectral line should be as sharp and narrow as possible, which means cooling the atoms. We now describe the different steps in the operation of a caesium fountain atomic clock.

Step 1 Laser cooling the atoms

Caesium atoms in the form of a gas are inserted into the vacuum chamber of the clock. A set of six infrared lasers firing their beams in both directions along orthogonal x, y and z directions is fired into the cloud of caesium atoms, which has the effect of pushing them into a ball and also cooling them. We shall consider the cooling by assuming that all motion is in one direction, so that caesium atoms can only move in the positive x (left-to-right) or negative x (right-to-left) directions.

To explain the cooling process, it is necessary to understand some fairly basic physics related to the concept of momentum. For a material body of mass m moving with a speed v, its momentum is given by

$$p = mv. \tag{4.3}$$

Thus, for a Cs-133 atom with mass 2.21×10^{-25} kg moving at 240 m s^{-1}, corresponding to a temperature of 300 K (27°C), the momentum is 5.30×10^{-23} kg m s^{-1}.

If a photon collides with an atom in such a way that its energy is absorbed by the atom, then its momentum is added to that of the atom. If the atom and photon are moving in opposite directions, then the magnitude of the momentum of the atom is reduced by the momentum of the photon, i.e. it slows down just as a moving body would be slowed down by a headwind. A photon of frequency v has energy hv (equation 4.1) and momentum

$$p_v = \frac{hv}{c}, \tag{4.4}$$

where c is the speed of light, 2.998×10^{8} m s^{-1}. For a photon with a typical infrared frequency, 3×10^{14} Hz, the momentum is 6.63×10^{-28} kg m s^{-1}. From this it can be seen that it would take the absorption of about 80,000 head-on collisions with infrared photons to bring a Cs-133 atom to rest.

Now we consider a collection of Cs-133 atoms, some moving in the positive x direction and others in the negative x direction. They are illuminated by two lasers from opposite directions with radiation of a frequency marginally below the level required to excite one of the atomic electrons from a lower to a higher energy level. Because of the Doppler effect, the effective energy of a photon when it strikes an atom is only equal to or higher than the excitation energy if the photon and atom are moving in opposite directions. In that case, the photon can be absorbed, promoting an electron to a higher energy state and giving its momentum to the atom, so slowing it down. However, the excited electron will fall back to its original energy level and emit a photon of similar momentum to the one absorbed. If it is emitted in the opposite

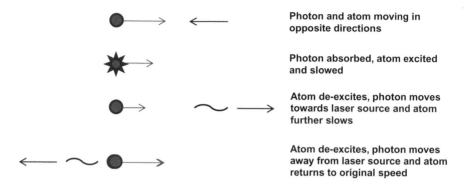

Figure 4.8 Possible outcomes of an atom absorbing a photon and subsequently de-exciting.

direction to that of the colliding photon, then it will slow the atom down further; if it is emitted in the same direction as the colliding photon, then it will neutralize the effect of the colliding photon and the speed of the atom will be restored to its original value. It will be seen that on average the effect of a collision is to slow down the atom. Photons colliding with the atom and moving in the same direction are not absorbed because their effective energy is too small to produce excitation. The effects of colliding photons in this one-dimensional example are shown in Figure 4.8.

Slowing down atoms is equivalent to lowering their temperature and temperatures as low as a fraction of a degree kelvin can be reached in this way within milliseconds. For the development of this technique, the Americans Steven Chu and William D. Phillips and the Frenchman Claude Cohen-Tannoudji were jointly awarded the Nobel Prize in Physics in 1997.

Step 2 Moving the ball and exciting the atoms

The intensities of the vertical lasers are adjusted to gently push the ball of caesium atoms upwards by about one metre through a microwave cavity, a partially enclosed container in which a microwave radiation field is set up. The set of six lasers is then switched off and the caesium ball falls back, again passing through the microwave cavity. The total journey time through the microwave cavity, there and back, is about one second, long enough for a substantial number of caesium atoms to be excited through the hyperfine transition, but only if the microwave frequency is appropriate. It is at this stage that the cooling process, which restricted the width of the spectral line that would be emitted, comes into its own. If the frequency of the microwave radiation is not within the width of the spectral line, then it will not excite electrons from the lower to the higher energy state corresponding to the caesium microwave emission frequency of interest. If it is just within the spread of the spectral line, then some of the caesium atoms will be excited. The closer the frequency of the radiation within the microwave cavity is to the centre of the caesium spectral line, the greater the number of excited caesium atoms. The microwave cavity can be tuned to change the frequency of the radiation field within it, so if the frequency can be found when the maximum number of caesium atoms is in an excited state, then it will be known that the frequency within the cavity is 9,192,631,770 Hz.

Stage 3 *Detecting the excited atoms*

When the caesium atoms have completed their double journey through the microwave cavity, the probe laser is switched on. This has a frequency such that the caesium atoms that have been excited can be further excited to a higher energy level, where the gain of energy corresponds to a frequency in the optical range. However, caesium atoms that have not been excited cannot be raised to that higher level because the photon energy associated with the frequency of the probe laser is insufficient (Figure 4.9). Those electrons that have been excited

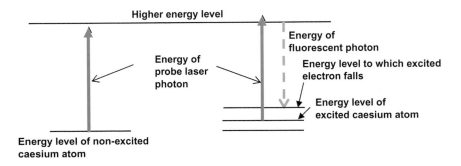

Figure 4.9   A photon from a probe laser moving only excited caesium atoms to a higher energy level.

then fall to a lower energy level, not the one from which they were excited, with the emission of a photon of visible light. This process, whereby an atom absorbs light and then emits light at a lower frequency is known as *fluorescence*. The total fluorescent light from the caesium atoms is measured by a detector and gives an indication of the closeness of the frequency of the radiation in the microwave cavity to that of the caesium standard. The frequency of the microwave cavity can be adjusted to give maximum fluorescence.

The frequency in the microwave cavity, now known to be 9,192,631,770 Hz, can be stabilized and maintained and the number of periods of the oscillations of the microwave radiation can be counted by electronic circuitry. The indication of time comes from this count. The use of microwave frequencies rather than optical frequencies in atomic clocks is to enable the electronic circuits to carry out the necessary counting operation At the very best, time is recorded with a precision of about 3 parts in 10^{16}, corresponding to an error of 1 second in 100 million years.

4.3.3 *Portable atomic clocks*

Caesium fountain clocks, as set up in the national standards institutes, are large and not portable, so atomic clocks that are smaller and portable, albeit less accurate, are in common use. One such clock is based on the hyperfine transition at 6,834,682,610.904324 Hz from the rubidium isotope Rb-87. The basic physics is the same as for the caesium fountain clock but implemented in a different way that allows for a much smaller and lightweight device. The basic operation of an Rb-87 clock can be followed by reference to Figure 4.9. Rubidium vapour, contained in a *resonance cell*, is subjected to microwave power whose frequency is controlled by a radiofrequency synthesizer, a device that can be tuned to a small range of frequencies spanning that of the hyperfine transition. A beam of light from a rubidium discharge lamp passes through the cell. If a photon encounters an unexcited rubidium atom, then it has insufficient energy to excite the atom to the higher energy level and so passes on without change. However, when it meets an excited atom, it has sufficient energy to excite it further and the photon is absorbed. The greater the concentration of rubidium atoms at the higher energy level of the hyperfine transition, the greater the absorption of light from the discharge lamp, which will be recorded as a lower intensity of light transmitted through the cell. The microwave frequency is swept through the hyperfine transition frequency of Rb-87 and when it hits that frequency, there is a maximum reduction of the intensity of the beam from the discharge lamp of about 0.1%. This is detected and an oscillator is automatically adjusted to the transition frequency and provides the means by which the passage of time is recorded.

An atomic clock of this kind can be as small as a matchbox and their prices are modest — less than 100 dollars, euros or pounds. The best rubidium clocks are accurate to about 1 part in 10^{12}, so in one year it could gain or lose about 30 microseconds, which is very little and a GPS satellite would only move about 24 centimetres in that time. Rubidium clocks are mounted in GPS satellites and act as *secondary frequency standards*. After a period of time, when the accumulated error is still acceptable, it receives a signal from a *primary frequency standard* — a caesium fountain clock — that resets it to zero, or virtually zero, error. If this is repeated periodically, then the satellite's atomic clock is always keeping time adequately well for GPS requirements, while only needing very occasional corrections.

Time and Relativity

Chapter 5

Time and Space

5.1 The Concept of Relativity

The word 'relativity' automatically brings to mind the work of the German (later American) scientist Albert Einstein (1879–1955), the imaginative 20th-century physicist who brought about a revolution in physics as great as that wrought by Isaac Newton (1642–1727) in the 17th century. However, the basic idea of relativity in the context of classical physics goes back to another scientific genius, Galileo Galilei, who was mentioned in Section 3.2 for originating the idea that a pendulum could be used to regulate the running of a clock.

Albert Einstein

Galileo's ideas of relativity are ones with which we are familiar in everyday life. Standing on a motorway bridge, you may see a car moving

75

at $110\,\mathrm{km\ hour}^{-1}$. If that vehicle overtakes another moving at $100\,\mathrm{km}$ hour^{-1}, then, as seen by the occupants of the slower car, the faster one is moving at a relative speed of $10\,\mathrm{km\ hour}^{-1}$. There is another observation that when sitting in a stationary train in a railway station you are under the impression that your train has resumed its journey, only to realize that you are stationary with respect to the station platform and that what you have seen is the neighbouring train setting off. Another common example is when driving while it rains: if it is not windy, then the rain may be falling vertically, but the impression in the car is that the rain is falling at a steep angle onto the windscreen. Relative to the car the rain has a horizontal component of motion with a speed equal and opposite to that of the car relative to the road. All these instances of classical relativity are well understood and agree with our understanding of the way things happen in the physical world.

There is another aspect of relativity that is equally obvious but is not often considered, which is that the laws of physics should be the same to all observers no matter how they are moving. It would indeed be odd if the laws of physics as deduced in a laboratory on Earth did not apply in a flying aircraft or to an astronaut standing on the Moon, which orbits the Earth and hence moves relative to it. If the laws of physics were not invariant, then we would have terrestrial physics, lunar physics and flying-aircraft physics, the last kind being of infinite variety because all aircraft are in motion relative to each other. Our common sense tells us that the science that applies on Earth also applies all over the Universe and indeed that is what is assumed in deducing the properties of distant galaxies by analysing the radiation that comes from them.

5.1.1 *Light and the ether*

There is a well-known experiment in which an electric bell rings continuously in a glass container attached to an air pump that slowly sucks out the air. As the pressure falls in the container, the sound of the bell gets progressively quieter and eventually, when there is a reasonable vacuum in the container, the bell cannot be heard at all. The clapper of the bell is seen to continue to vibrate but the sound waves are unable to reach the glass walls of the container and thence the outside world.

This shows that the air was needed to transmit the sound and in general material is necessary to transmit sound waves.

It was believed up to the first decade or so of the 20th century that any kind of wave motion requires a material for it to be transmitted, and electromagnetic waves were not excluded from this requirement. It was known that light could travel to the Earth from far reaches of the Universe across what was regarded as the vacuum between galaxies, so the problem was to identify the medium in which the light was moving. The space between galaxies contains *some* material, but with an expected density such that there would be several centimetres between one atom and the next. As far as visible light is concerned, with its wavelength less than one micron,[1] intergalactic space — and even interstellar space within a galaxy — behaves like a perfect vacuum.

The answer to this conundrum was to postulate that the whole of the Universe was pervaded by a medium, *the ether*, within which light travelled. The ether had density so low that it could not be directly detected and high elasticity, meaning that it was very stiff, a combination of properties that explained the high speed of light. This was just a postulate, but physicists began to think of experiments by which they could detect the ether by the way that it affected light travelling through it.

5.1.2 *An attempt to detect the ether*

If the ether permeated the whole of space, then it provided an absolute standard for the state of being at rest, because the motion of all bodies could be related to the stationary ether. In 1887, two American scientists — Albert Michelson (1852–1931), a physicist awarded the Nobel Prize for Physics in 1907, and the chemist Edward Morley (1838–1923) — carried out an experiment to detect the motion of the Earth though the ether. To explain the basis of this experiment we shall first consider an analogous experiment in terms of objects with properties more familiar to us than those of light.

---

[1]A micron (μm) is one-millionth of a metre.

Albert Michelson   Edward Morley

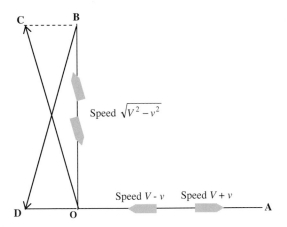

Figure 5.1 The boat race.

A boat race

In Figure 5.1 there are two similar boats situated at point O. Points A and B are equidistant, at distance D, from O, with the water flowing with a speed v along the direction OA. Now, if the boats set off at speed V relative to the water, which is much greater than v, one on a return trip OAO and the other on a return trip OBO, then which will be the first to return to O? Here the question will be answered by a mathematical approach, but if mathematics isn't your thing, then just skip to the end — the conclusion is what matters.

A mathematical interlude

The journey from O to A is with the flow of the water so the speed of the boat along OA on this leg of the journey is $V + v$. However, the return journey is against the flow so the speed of the boat along AO is $V - v$. Hence the time for the return journey is

$$t_{OAO} = \frac{D}{V + v} + \frac{D}{V - v} = \frac{2DV}{V^2 - v^2}. \tag{5.1}$$

Considering the journey from O to B and back across the flow of water is more complicated. If the boat were to steer towards B, it would not reach B, since it would be swept to the right by the flow of water. To reach B it must point the bow in the direction OC where CB/OC $= v/V$. Treating the distances CB and OC as v and V, the actual speed of progress along OB is given by Pythagoras' theorem as

$$V' = \sqrt{V^2 - v^2}.$$

Similarly, on the return journey the boat must point along the direction BD to reach O and again the speed along BO is V'. Thus, the return journey OBO takes time

$$t_{OBO} = \frac{2D}{V'} = \frac{2D}{\sqrt{V^2 - v^2}}. \tag{5.2}$$

It is clear that the journey times t_{OAO} and t_{OBO} are not the same. In the situation where v^2 is much less than V^2, the difference between the two times can be found by a special application of the binomial theorem, viz.

$$(1 + \varepsilon)^n \approx 1 + n\varepsilon, \tag{5.3}$$

when ε is so small that ε^2 and higher powers of ε can be ignored. Using this approximation

$$t_{OAO} = \frac{2DV}{V^2 - v^2} = \frac{2D}{V}\left(1 - \frac{v^2}{V^2}\right)^{-1} = \frac{2D}{V}\left(1 + \frac{v^2}{V^2}\right)$$

and

$$t_{OBO} = \frac{2D}{V}\left(1 - \frac{v^2}{V^2}\right)^{-1/2} = \frac{2D}{V}\left(1 + \frac{v^2}{2V^2}\right),$$

hence

$$t_{OAO} - t_{OBO} = \frac{Dv^2}{V^3}. \tag{5.4}$$

Conclusion

The time for the OAO journey, t_{OAO}, is greater than that for the OBO journey, t_{OBO}, by an amount Dv^2/V^3.

The Michelson–Morley experiment

We now consider the Michelson–Morley experiment, which is a precise analogue of the boat race and is shown in Figure 5.2. Light coming from the source in the form of a fine beam is split by a half-silvered mirror, the part transmitted going along OA and the part reflected going along OB. The light striking the mirror at A is reflected back along its approach path and then is reflected by the half-silvered mirror into the detector. The light reflected from the mirror at B is transmitted

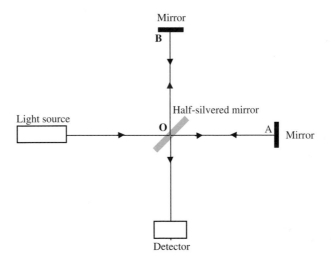

Figure 5.2   A schematic of the Michelson–Morley experiment.

Figure 5.3   Fringes in the detector from the Michelson–Morley experiment.

by the half-silvered mirror to enter the detector. The distances OA and OB are made to be the same, or very closely so, and the beams of light entering the detector form an interference pattern as shown in Figure 5.3.

Assuming that OA and OB are the same length, then the paths of the two beams are equal. However, if the motion of the Earth relative to the ether is at speed v along the negative x direction, then the ether is moving at speed v along the positive x direction relative to the apparatus — this is equivalent to the flow speed of the water for the boat race. From the analogy of the boats example the light from the path OAO will take more time by an amount Dv^2/c^3, where c, the speed of light, 300,000 km s^{-1}, is equivalent to the speed of the boats. If the light frequency is f corresponding to an oscillation period $\tau = 1/f$, then the waves from the OAO path will be behind those from the OBO path by a fraction

$$\phi = Dv^2/(c^3\tau) = Dv^2f/c^3 \qquad (5.5)$$

of the wavelength of the light. To get an idea of the magnitude, if $D = 5$ m, $v = 50$ km s^{-1} and $f = 5 \times 10^{14}$ s^{-1}, corresponding to an orange wavelength of light, then $\phi = 1.2$. The next step is to rotate the apparatus through a right angle so that OB is now parallel to the relative motion of the ether. Now it is the beam OBO which lags behind OAO and consequently there should be a change in the interference pattern seen in the detector by a shift of 2.4 fringes, using the above parameters for the experiment. No change was seen.

It is possible that, just by chance, the Earth was stationary with respect to the ether at the time of the experiment. Since the Earth orbits the Sun at a speed of about $30 \, \text{km} \, \text{s}^{-1}$, then after six months it will have changed its velocity with respect to the ether by $60 \, \text{km} \, \text{s}^{-1}$; if it happened to be at rest with respect to the ether on the first occasion, then six months later the experiment should give a positive and easily measurable result. But again, there was no change in the interference pattern when the experiment was repeated. The attempt to detect the ether had failed and raised the question of whether or not the ether existed.

5.2 Einstein's Postulate

The negative outcome of the Michelson–Morley experiment presented a conundrum to the scientific community. The 60 kilometres per second difference in the speed of the Earth with respect to the ether after an interval of six months should have given an easily measurable effect. One suggestion was that the Earth dragged the ether along with it in its immediate vicinity, so that no matter where the Earth was in its orbit it was always stationary with respect to the ether that enveloped it. This was not a very satisfactory explanation and was contradictory to the original concept of the ether as providing an absolute standard of rest against which all motion could be measured.

The answer to the problem was provided by Albert Einstein who solved it very simply by making the postulate that the speed of light was a constant to all observers. We have previously mentioned the proposition that the laws of physics have to be the same to all observers and what Einstein's postulate did was to add the constancy of the speed of light as one of the laws of physics. However, this postulate had implications for the way that we regard time and space, something that we now discuss. It also had more practical implications in the way that it related mass and energy; under the right conditions each of them could be changed into the other. Mass-to-energy conversion is the basis of the production of nuclear energy and also of nuclear weapons.

5.2.1 *Physics at the beginning of the 20th century*

We all have an instinctive understanding of space and time. Space has three dimensions specified in a non-scientific way as forwards–backwards, sideways and up–down. Time is something we record in various ways with varying degrees of accuracy as we have previously described. In 1687, Isaac Newton expressed in his great work *Philosophiæ Naturalis Principia Mathematica* (commonly known as *Principia*), in the somewhat poetic language of his day, what we instinctively understand about the nature of time and space:

- 'Absolute, true, and mathematical time, from its own nature, passes equably without relation to anything external.'
- 'Absolute, true, and mathematical space remains similar and immovable without relation to anything external.'

Isaac Newton

According to these instinctive ideas of space and time, the place and time of any event will be the same to all observers. This is what our experience tells us is true and we are comfortable with those things that conform with our experience. We do not need a scientist to tell us that light travels in a straight line and that we cannot see around corners. A child tossing a ball to a friend knows that it will move in the form of an arc first rising up and then descending; she knows nothing about mechanics but her experience tells her what will happen. There is survival value in accepting what experience tells us is true. However, we have no experience of small entities such as atoms or large ones such as black holes. The bodies seen in everyday life move at speeds that are tiny compared with the speed of light. In considering these matters, our experience is non-existent and we must turn to the scientist for guidance as to how such things behave.

At the beginning of the 20th century, the physical world seemed to be divided into two categories of phenomena that were described by two apparently unconnected theoretical bases. Newtonian mechanics described the behaviour of solid objects and included the effect of gravity that exerted force at a distance without the need for intervening material. The second category of theories, less well known but no less important, was due to the work of the Scottish physicist James Clerk Maxwell (1831–1879). Despite his short life, Maxwell contributed to many topics in physics, in particular by explaining the interactions between electricity and magnetism and by showing that light is an electromagnetic wave. The two basic and apparently independent theories provided by Newton and Maxwell seemed together to deal with all aspects of physics.

James Clerk Maxwell

During the early years of the 20th century, experiments were being done that connected the two sets of phenomena described by the theories of Newton and Maxwell. Electrons, normally thought of as negatively-charged particles, could sometimes behave like light — wave-like electromagnetic radiation. Conversely, light, normally thought of as wave-like electromagnetic radiation, could sometimes behave like particles, with the energy of the light existing in bullet-like entities called *photons*. The behaviour of electrons and light depended on the experiment being performed; an experiment designed to detect light as photons would do so, and in an electron microscope the electrons would form an image, just as ordinary light would do with a conventional microscope.

This describes the state of physics when Einstein began to explore the consequence of his proposal that the speed of light is the same to all observers.

5.3 The Theories of Relativity and their Consequences

The apparently bland assumption that the speed of light is the same to all observers was anything but bland — it changed the world of science in a profound way and to some extent changed our understanding of physical science away from that provided by Newton 300 years earlier. It led in 1905 to what is known as Einstein's *Theory of Special Relativity*, the conclusions of which are completely in conflict with the Newtonian world of our experience.

Einstein's world is one that is counter-intuitive and very strange. Someone in a train travelling in a straight line from London to Edinburgh at 200 km per hour would assess the length of the journey as 535 km. If he repeated the journey at one-half of the speed of light, the distance would be assessed as 463 km. There is a very simple equation that links the assessed distance at high speed d_α, some considerable fraction α, of the speed of light, and the distance that would be measured by a slow-moving individual d_0. This relationship is

$$d_\alpha = d_0\sqrt{1 - \alpha^2} \qquad (5.6)$$

so that, for example, with $d_0 = 535$ km and $\alpha = 0.5$ (half the speed of light) the fast traveller records the distance as

$$d_{1/2} = 535\sqrt{1 - \left(\frac{1}{2}\right)^2} = 463 \text{ km.}$$

There is another unexpected effect to do with the recording of time. An observer standing by the trackside outside the fast-moving train looking at a clock within the train would notice that it was running rather slowly, taking 69.3 seconds for the hands to move forward 1 minute, as seen on the observer's watch. The distances and times measured by the person on the train and the one standing by the track would be different. The person by the trackside would, by any experiment he could do, find the distance from London to Edinburgh to be 535 km while the one on the train would find that the clock on the train was keeping perfect time by comparison with his watch or by comparing the running of the clock with some physiological process such as the rate of his pulse. Another odd observation made by the trackside observer is that the train, and everything in it, would seem strangely compressed. London to Edinburgh trains are a standard length, 150 metres, but to the trackside observer the train he saw would seem only to be about 130 metres long (Figure 5.4). The shortening fraction is the $\sqrt{1 - \alpha^2}$ we have already met.

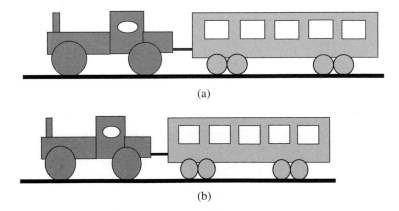

(a)

(b)

Figure 5.4   (a) The train as seen moving slowly. (b) The train as seen moving at half the speed of light.

You might say, 'It is all very well to just quote these relationships, but is there any way that they can be verified?' Yes, there are plenty of ways, but we will just mention two of them here. When we say that the minute hand of the clock on the train was taking 69.3 seconds to record 1 minute as seen by the observer by the trackside, we say that *a moving clock runs slowly*. We might notice our old friend $\sqrt{1 - \alpha^2}$ appearing again, because $60 \div \sqrt{1 - \left(\frac{1}{2}\right)^2} = 69.3$. There are some elementary charged particles called *mesons* that decay into something else at a certain rate. These particles can be accelerated in a particle accelerator to travel on a circular path at a very high speed — very close to the speed of light — and the mesons are seen to decay much more slowly. The decay of mesons can be considered as a way of measuring time — a kind of clock — and the moving meson clock runs slowly. Another similar test can be made using a very precise atomic clock mounted in an Earth satellite moving around the Earth at 8 km s^{-1}. For this experiment the value of α is given by

$$\alpha = \frac{\text{speed of satellite}}{\text{speed of light}} = \frac{8}{300,000} = 2.6667 \times 10^{-5}.$$

If we apply this value of α to our moving atomic clock, we find that it runs more slowly than a clock on Earth by 30 microseconds per day, a delay that an atomic clock can record quite accurately, especially over a longer period.

A very strange conclusion from Einstein's theory, one that our instinct strongly rebels against, is the *twin paradox*. One twin travels by spacecraft to and from a nearby star at three-quarters of the speed of light. He returns after 30 years according to the spacecraft clock. On his return he finds that to people on Earth his journey took 45 years and when he meets his twin who remained on Earth, he sees that indeed his twin has aged 15 years more than he has (Figure 5.5). This is equivalent to the slow-running clock phenomenon. Ageing is a kind of time-measuring device and the twin that travelled was 'running slowly'.

You may have spotted an apparent flaw in this narrative — after all, if all motion is relative, then since the Earth-bound twin was moving at three-quarters of the speed of light *relative to his travelling brother,*

Cheerio, Joe – have a good trip.

Thanks, Fred – see you sometime.

AD 2020

Welcome home, Joe – you're looking well.

Is that you, Fred?

AD 2065

Figure 5.5   The twin paradox. The traveller comes home younger than his twin brother.

why is the age difference not the other way round? It is an argument that could also have been used for the atomic clock in the spacecraft — why were it not the clocks on Earth that were running slowly? The explanation is that in order to leave the Earth and then return to Earth, the travelling twin underwent accelerations; similarly, a satellite in Earth orbit is in a state of constant acceleration in order for it to move on a curved path. For accelerating bodies or those experiencing a force, e.g. those close to a massive body, another one of Einstein's theories applies, *The Theory of General Relativity*, introduced in 1915, and applying this resolves the apparent paradox in the application of the Special Theory. During the acceleration parts of the traveller's journey, if he could have seen his brother on Earth, he would have seen him ageing at a greater rate than he was. This would have more than compensated for the effect during the constant speed part of his journey where his brother would have appeared to be ageing more slowly and restored the observed asymmetry to the ageing process.

5.3.1 *Spacetime*

In our everyday lives we are accustomed to an understanding of space in three dimensions quite separate from time, which is of a completely different nature and has the characteristic that it can only flow one way — from the past to the present to the future. In Einstein's world, space and time are intimately mixed and the composite of the two is referred to as *spacetime*. Scientists are accustomed to changing coordinate systems to describe the position of an object; this might mean just changing one set of coordinates where the position of an object is given by (x, y, z) to another where the coordinates are (X, Y, Z), the two coordinate systems having origins displaced from each other and with the lowercase and uppercase axes not parallel to each other. In the relativistic world, there are four coordinates, with time added to space, and the two sets of coordinates can be described as (x, y, z, t) and (X, Y, Z, T). The progress of any event, for example the motion of an object, can be plotted in spacetime, a four-dimensional space which indicates where an object is situated and when it was there. Four dimensions cannot be depicted graphically and on a plane surface we can only depict two dimensions. In Figure 5.6 we show the motion of an object moving in one space dimension and the time when it was at each position.

In Figure 5.6a the plot is a straight line and equal intervals of apace always correspond to equal intervals of time, which indicates motion

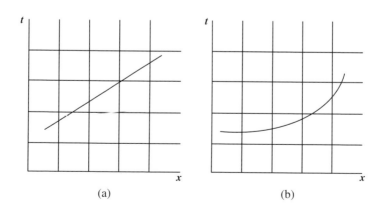

(a)   (b)

Figure 5.6 A spacetime plot for (a) uniform motion and (b) accelerated motion.

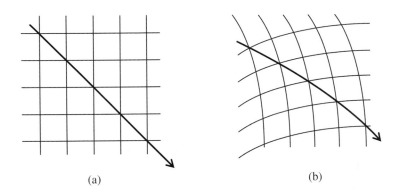

(a)   (b)

Figure 5.7 (a) A grid defining position in a two-dimensional space and the path of a particle in the space. (b) A distortion of the grid due to a gravitational field with corresponding distortion of the path of the particle.

at a uniform speed. By contrast, Figure 5.6b shows a curved line and equal intervals of space correspond to unequal intervals of time, which is an indication of accelerated (or decelerated) motion. In the latter case, for a change of velocity to occur, the body must be experiencing a force.

Special Relativity only applies to bodies that are not experiencing forces and hence do not accelerate. Einstein's Theory of General Relativity deals with accelerating bodies and the forces that cause acceleration such as gravity. According to this theory, the presence of a massive body distorts spacetime, so causing the body to deviate from a straight path corresponding to uniform velocity. Figure 5.7a shows a two-dimensional grid with a particle moving in a straight line. The presence of a massive body distorts this grid, as shown in Figure 5.7b, and consequently the path of the body is distorted. We must imagine this distortion taking place in the four-dimensional world of spacetime — a difficult feat for human imagination but straightforward to express mathematically.

5.3.2 *Time and gravity*

Another interesting consequence of Einstein's Theory of General Relativity is that clocks close to a gravitational source, i.e. a massive body, will run more slowly than they would if well away from such a source. The effect is a small one, but it has been verified experimentally. According to the theory, if a clock at the Earth's surface were compared with another clock at a great distance from the Earth and not close to any other massive body then it would seem to be losing 60 microseconds per day, or 0.0219 seconds per year. If the body were the Sun rather than the Earth, then the clock at the Sun's surface would lose 66.4 seconds per year.

Although the effect is very small, it can be detected by experiments done on Earth. If an atomic clock at sea level is compared with one at an altitude of 4 kilometres, placing it further from the centre of the Earth, it would be found that the higher clock would be gaining 18.8 nanoseconds per day relative to the lower one, or 0.13 microseconds per week, easily within the measurement capability of atomic clocks. The effect is a real one, of interest to scientists, if not relevant to everyday life.

5.3.3 *Before, after or simultaneously?*

In Newton's picture of time, described in Section 5.2.1, it flows smoothly in one direction, marked out by a series of events that occur in a sequence that would be common to all observers. Sometimes, events would occur at the same time — they would be *simultaneous.* However, the very first word of Newton's description of the nature of time — 'absolute' — is shown to be in error by relativity theory; simultaneity, or even the order in which events occur, is not absolute but depends on the observer and the relationship of his reference frame to the one in which the events occur.

A very simple thought experiment illustrates the way in which the order of events may be seen by different observers. In Figure 5.8a we see a moving railway carriage with an observer at its midpoint. The carriage is in darkness and an extremely brief pulse of light is emitted at the midpoint of the carriage and travels along the carriage in both directions and, as expected, the observer sees the ends of the carriages illuminated simultaneously. Now we consider what would be seen by an observer standing on a platform in such a position that he was level with the midpoint of the carriage when the pulse of light was emitted. According to Einstein's postulate, this observer sees both pulses moving at the speed of light, so the situation as seen by this observer is illustrated

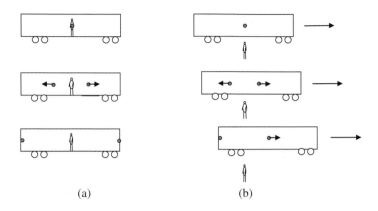

(a)   (b)

Figure 5.8   (a) To the observer in the carriage, the light pulses reach the ends of
the carriage simultaneously. (b) To the observer on the platform, the
light pulse reaches the rear of the carriage before the front.

in Figure 5.8b. Now the pulse of light moving backwards is being
approached by the rear of the carriage, while the front of the carriage
is moving in the same direction as the light pulse. Consequently, the
observer on the platform sees the rear of the carriage illuminated before
the front. Since all motion is relative, then if the observer were travelling
in the same direction as the train but at a greater speed, in a relative
sense the train would be moving in the reverse direction. Then that
observer would see the front of the carriage illuminated before the
rear. So two events, the light arriving at the front of the carriage and at
the rear, may be seen as simultaneous or one after the other in either
order, depending on the observer.

However, although the order in which events are seen can be
different in some circumstances, there are limitations. For example,
you cannot see someone being felled by a cricket ball and then later
see the batsman hit the ball that caused the injury. Where events have
a causal relationship so that the occurrence of one is a consequence of
the other, then they can only be seen in the order that our common
sense dictates. This principle can be made more general. If when the
first event occurs a pulse of light is emitted and reaches the location
of the second event before it happens, then, in principle, the second
event could have been caused by the first — for example by the arriving

light activating a switch that caused it to happen. In such a case the first event must be seen before the second by any observer. Conversely, if the pulse of light does not reach the location of the second event before it happens, then there can be no causal relationship between the events and different observers can see the events in either order or simultaneously.

Einstein's world is full of surprises — space and time and energy and mass are curiously related in ways that our intuitive understanding of the world finds very strange.

The Ages of Astronomical Entities

Chapter 6

The Age of the Universe

6.1 Observing the Universe

Looking out on the night sky, one sees a myriad of stars, some shining brightly but others quite faint. Those people living in remote areas, well away from the artificial light of urban life, see far more than those in cities. However, it required the advent of fairly advanced telescopes to see objects in the sky discernable as something other than individual stars. The astronomer William Herschel (1738–1822), who discovered the planet Uranus in 1781, built a groundbreaking telescope, the '40-foot reflector', in the garden of his home in Bath and with it he observed hazy patches of light in the sky that he speculated were 'island universes', the first suggestion of the existence of distant galaxies. Now we know that the stars that we see with the naked eye are denizens of our own home galaxy, the Milky Way, and that there are other galaxies, well separated from our own, some much bigger than ours and others much smaller. With the development of more powerful telescopes, galaxies have been observed at ever greater distances, as judged by their decreasing angular size and the reduction in their brightness. The question that astronomers were asking is how far away these galaxies are and another question, of a more philosophical nature, was whether or not an age could be assigned to the Universe — the totality of what could be observed. The question was whether the Universe had always existed so that time, the measurer of change, had no beginning or whether there was an instant that could be regarded as the beginning of the Universe and the beginning of time.

The problem of determining the distances of astronomical objects was what first engaged the energies of astronomers and techniques were developed progressively to extend the distances that could be determined. This was a necessary forerunner of the method that was eventually used to determine an age of the Universe, so now we give a brief summary of those techniques.

6.2 The Distances of Astronomical Objects

The history of the measurement of the distances of astronomical objects is based on the principle that the first measurements were of close objects — nearby stars — and subsequent techniques extended the range of the distances that could be measured but depended on being anchored to, and calibrated by, distances measured by earlier techniques.

6.2.1 *The parallax method*

If you hold a finger vertically at arm's length and alternately close one eye and then the other, the finger will be seen to move against a distant background. If you bring your finger closer to your face and repeat the observations, you will find that the movement of the finger against the background is greater. This phenomenon is known as *parallax* and the *parallax method* is used to measure the distance of nearby stars, where nearby means being within a few hundred light years.[1]

The principle of the method is shown in Figure 6.1. Points A and B are the positions of the Earth in its orbit around the Sun, providing two viewing positions. When at position A, the line of sight towards the star is noted on the background of very distant stars, which are thousands of times more distant than the star of interest and maintain the same directions in space over historical time. The same is done six months later from position B. Now, the distance from A to B, the diameter of the Earth's orbit, is known,[2] so that if the angle α can be found,

---

[1] The light year (ly) is the distance that light travels in 1 year, 9.461×10^{15} m.

[2] This distance is two astronomical units (au), where 1 au is the mean Sun–Earth distance, 1.496×10^{11} m.

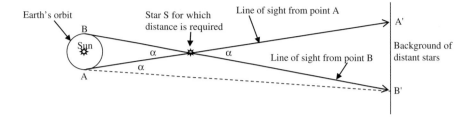

Figure 6.1 The basis of the parallax method.

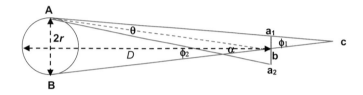

Figure 6.2 The three observations of a star to find its distance and speed perpendicular to the line of sight.

then the distance of the star could be determined. Because of the great distance of the background of stars, the angle A'AB' is also α, with a tiny immeasurable error, and this angle, usually in the order of a second of arc, can be measured.

Finding the distance and lateral speed of a star

There is the problem that in the interval between taking the readings at A and B, the star would move, but by taking measurements at A, B and then A again not only can the distance of the star be found but also its speed relative to the Sun perpendicular to the line of sight.

A mathematical interlude

Figure 6.2 shows observations from points A, B and A again in the Earth's orbit with the star moving across the field of view at corresponding points a_1, b and a_2. The lines of sight are Aa_1, Bb and Aa_2 and the angles between them are indicated in the figure.

To determine the distance, the angle required is α between Ab and Bb, but Ab is not one of the observations. However, if we consider the triangle Abc then α is the external angle at b and equals the sum of the other two internal angles. The angle at c is ϕ_1 and that at A is $\theta/2$ since Ab bisects the angle θ. Hence

$$\alpha = \phi_1 + \theta/2. \tag{6.1a}$$

An independent estimate of α can be made from

$$\alpha = \phi_2 - \theta/2 \tag{6.1b}$$

and an average of the two estimates can be used.

From the value of α, the distance of the star is found from

$$D = 2r/\alpha, \tag{6.2}$$

where r is the radius of the Earth's orbit.

The star moves a distance $a_1 a_2$ $(= D\theta)$ perpendicular to the line of sight in a period of one year so its transverse speed is

$$v_t = D\theta/t_y, \tag{6.3}$$

where t_y is one year.

Conclusion

Making three observations of the star from points A, B and A again gives both the distance of the star and its transverse speed. The radial speed — the component of velocity along the line of sight — can be found from Doppler-shift measurements (Section 6.2.4). The radial movement in one year is trivial compared with the distance of the star and makes no difference to the mathematical analysis.

With ground-based telescopes that use *adaptive optics*[3] to remove the shimmering effect when light moves through the turbulent atmosphere, distances out to about 200 ly can be measured. The Hipparchos satellite, launched by the European Space Agency in 1989 specifically to make parallax measurements, has enabled the distances of about 1,000,000 stars to be measured out to a distance of about 600 ly. Although this may seem a large number it is only one-hundred-thousandth of the stars in the Milky Way galaxy.

---

[3] Details of adaptive optics will be found in the author's book *The Fundamentals of Imaging: From Particles to Galaxies*, 2012, pp. 158–167 (London: Imperial College Press).

6.2.2 *Main-sequence stars*

The Sun is a *main-sequence star*, meaning that nuclear reactions in its core are producing energy by converting hydrogen to helium. Eventually, when the hydrogen in the core is depleted, the star evolves through a number of stages with the kind of final body — white dwarf, neutron star or black hole — depending on the mass of the star. Here we are just concerned with the main-sequence stage of a star's existence.

The Sun is about halfway through its ten-billion-year lifetime as a main-sequence star. During this period its mass, radius and temperature will remain approximately constant. Relative to the other stages of the existence of a star as a bright object, the main sequence is long-lasting so that a high proportion of the stars we observe are main-sequence stars. For many of the stars within the parallax range, it is possible to determine their mass, temperature and *luminosity*, the total energy output from the star that governs its brightness.

There are as many *binary star* systems, where two stars orbit around each other, as there are single isolated stars, known as *field stars*. In some binary systems — *visual binaries* — the two stars are well separated and may be seen individually. However, for most binary systems the individual cannot be resolved and for these *spectroscopic binaries* the fact that there are two stars present must be deduced from Doppler shifts in the wavelengths of the spectral lines they emit. By observations of the orbiting motions of stars in visual binaries it is possible to find their individual masses.[4]

The temperature of a star is revealed by its colour — red stars are comparatively cool while bluish-white stars are very hot. Actually, astronomers estimate the temperature of a main-sequence star not by looking at the wavelengths it emits but rather by looking at the *Fraunhofer absorption lines* seen in Figure 6.3. These lines are due to the absorption of specific wavelengths of light by the atoms present in the outer layers of the star and they individually vary in intensity in different ways as the temperature changes. The principle of this is

---

[4]Details of this process will be found in the author's book, *On the Origin of Planets: By Simple Natural Processes*, 2011, pp. 338–340 (London: Imperial College Press).

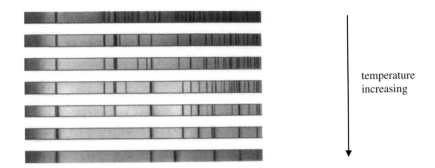

temperature
increasing

Figure 6.3   The change in the pattern of spectral lines with temperature. The lowest temperature (at the top) is about 3,000°C and the highest temperature (at the bottom) about 30,000°C.

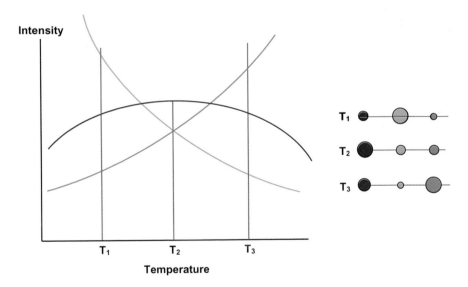

Figure 6.4   The variation of the intensity of three spectral lines with temperature.

illustrated in Figure 6.4, which shows how the variation in the relative intensity of three spectral lines — one red, one green and one blue — can give an indication of temperature.

The luminosity of a star is defined as its total energy output per unit of time and can be found from a combination of its apparent brightness, as seen from Earth, and its known distance. The brightness of an object is inversely proportional to the square of its distance — e.g. doubling the distance of the star reduces its brightness by a factor of four. Theoretically, it can be shown that the luminosity L of a star depends on its surface temperature T and its radius R through the relationship

$$L = 4\pi\sigma R^2 T^4, \tag{6.4}$$

in which σ is Stefan's constant, $5.670 \times 10^{-8}\,\mathrm{W\,m^{-2}\,K^{-4}}$. From any two measurements of L, T and R the third quantity can be found from equation (6.4). For example, the Sun has a radius of $6.960 \times 10^8\,\mathrm{m}$ and temperature 5,778 K. Hence, from (6.4) its luminosity is $3.85 \times 10^{26}\,\mathrm{W}$.[5] This power is provided by the mass loss when hydrogen is converted to helium by nuclear reactions in the Sun's core; the mass loss per second, according to Einstein's famous equation linking mass and energy, is $m = L/c^2 = 4.3 \times 10^9\,\mathrm{kg\,s^{-1}}$ — more than four million tonnes per second! This need not concern us since it represents a loss of only 0.007% of the mass of the Sun in one billion years.

6.2.3 *Using Cepheid variables*

The pioneer in the study of variable stars, the brightness of which varies in a periodic way, was the English astronomer, John Goodricke (1764–1786). Despite the social handicap in those days of being profoundly deaf, he acquired an education and became an astronomer.

A significant discovery by Goodricke was the variable star δ-Cephei, which fluctuates in brightness with a period of about 5.4 days; Goodricke's early death, at the age of 21, was probably due to pneumonia, contracted as a result of exposure when observing δ-Cephei. This star was the prototype of stars called *Cepheid variables* which differ in their average luminosity and period. Many of these stars are within the parallax range so that their distances and hence their average luminosity can be found.

---

[5]The watt (W) is a power of 1 joule per second.

John Goodricke

In 1908, the Harvard astronomer Henrietta Swan Leavitt (1868–1921) found a relationship between the average luminosity of Cepheid variables and their periods (Figure 6.5). Cepheid variables can be more than 30,000 times brighter than the Sun, and so can be seen at great distances; in particular those that occur in the outer regions of some nearer galaxies. From the results in Figure 6.5, a measurement of the period of a Cepheid variable in a galaxy gives its average luminosity and then, from the measured apparent average brightness, the distance of the galaxy can be found. This technique enables distances out to about 80 million light years to be estimated.

Henrietta Swan Leavitt

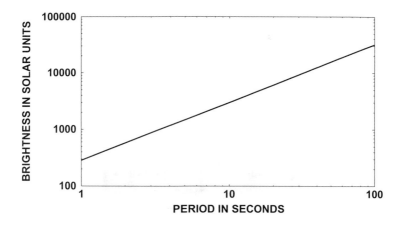

Figure 6.5 The relationship between the average luminosity of a Cepheid variable star and its period.

6.2.4 *The Doppler effect*

The Doppler effect, introduced in Section 4.3.2, which applies to any kind of wave motion, was first described by Austrian mathematician Christian Doppler (1803–1853). Any wave motion is characterized by three parameters — frequency, ν, wavelength, λ, and speed, c — which are related by

$$\lambda \nu = c. \tag{6.5}$$

Christian Doppler

The Doppler effect gives a change in the detected wavelength (or frequency) from wave sources that are either approaching or receding from the observer. If the speed of the source along the line that joins source and observer is V then the change in the wavelength detected is given by

$$\frac{d\lambda}{\lambda} = \frac{V}{c},\qquad(6.6)$$

in which V is positive if the source is moving away from the observer and negative otherwise. For visible light the longer wavelengths are at the red end of the spectrum, so that if a light-emitting body is moving away from the observer, then V is positive and so is $d\lambda$, meaning that the wavelength moves in the direction of the red end of the spectrum. The light is said to be *red shifted*. Alternatively, if the source is moving towards the observer, then the observed wavelength is reduced and the light is said to be *blue shifted*. The Doppler effect is a very important astronomical tool and now we show one way in which it can lead to distance estimation.

6.2.5 *Spinning galaxies*

To extend estimates of distance beyond the Cepheid variable range a much brighter object is needed and the ones we use are spiral galaxies, similar to our own Milky Way (Figure 6.6). From determining the motions of stars within the Milky Way we know that our galaxy is spinning, making a complete revolution in 200 million years. That is a slow rotation but, because of the immense size of the galaxy with a diameter more than 100,000 ly, stars on opposite sides of the galaxy have a relative speed of about 1,000 kilometres per second. Someone observing the Milky Way edge-on would see the stars on one side of the galaxy moving away relative to the centre of the galaxy while those on the other side were approaching. Assuming that the centre of the galaxy was at rest with respect to the observer, there would be a red shift for the stars on one side and a blue shift for stars on the other side (Figure 6.7). The light would not be seen from individual stars but as the aggregate light from all stars in different regions of the galaxy. The light from the centre would have no shift and as one looked further in

Figure 6.6 The Pinwheel Galaxy, similar to the Milky Way.

Moving towards
–blue shifted

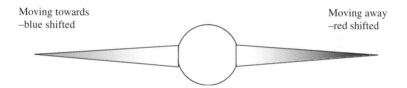

Moving away
–red shifted

Figure 6.7 The spectral line shifts from a rotating spiral galaxy seen edge-on.

one direction the red shift would steadily increase and there would be a similar increase in blue shift as one looked further in the other direction. A spectral line of the kind shown in Figure 6.3 would be spread out due to the different wavelength shifts from different parts of the galaxy and the width of the spread would give a measure of the rotational speed of the galaxy. This spreading effect would not be affected by the overall motion of the galaxy towards or away from the observer. The centre of the line would be red shifted or blue shifted but the width of the line would be unaffected.

It was found that for spiral galaxies within the Cepheid variable range the width of spectral lines is closely related to the total luminosity of the galaxy. The physical link between luminosity and rotation speed is the galactic mass. The mechanics of a spiral galaxy give a rotation speed increasing with total mass. However, the number of stars in, and hence the brightness of, a galaxy also increases with the galactic mass; thus faster rotation is correlated with greater brightness, the connection between these two quantities being known as the *Tully–Fisher relationship.* By measuring the spread of the spectral lines of a distant spiral galaxy its luminosity can be estimated, and its apparent brightness then gives its distance. Using this method, distances out to about 600 million light years can be measured, beyond which the images of galaxies are too faint to give a good estimate of the spectral-line widths.

6.2.6 *Using supernovae as standard sources*

At the end of the Sun's main-sequence period it will shed outer material in the form of a planetary nebula (Figure 6.8), eventually leaving behind an extremely dense core in the form of a *white dwarf* (Section 7.3.2), an object of Earth's size but with about the mass of the Sun. A teaspoonful of white-dwarf material would have a mass of about 10 tonnes! A star much heavier than the Sun would undergo a more violent end,

Figure 6.8   A planetary nebula — a complete shell of material although, in projection, it looks like a ring.

completing its main-sequence existence in a violent *supernova* explosion when for some weeks it can be as bright as ten billion Suns.

There are other kinds of supernovae, which can be individually recognized by features of their spectra. One such supernova occurs when a white dwarf is in a binary system with a kind of star called a *red giant* (Section 7.3.2), which is comparatively cool, but very large and in an intermediate stage between being on the main sequence and becoming a white dwarf. Red giants shed material and if it is in a binary system with a white dwarf then some of the shed material will attach itself to the white dwarf. A white dwarf in isolation will gradually cool down until it ceases to shine, at which stage it becomes a *black dwarf*. However, if it gains mass while it is a white dwarf, it can reach a critical mass known as the *Chandrasekhar limit*, about 1.44 times the mass of the Sun, at which stage it becomes unstable and explodes to give what is known as a *type 1a supernova*. Because all type 1a supernovae come about in the same way — they all reach the same critical mass — they are very similar and in particular the peak luminosity is the same from one type 1a supernova to another.

Type 1a supernovae are very bright and can be seen in distant galaxies. Because some have occurred within galaxies for which Cepheid-variable distances are available their maximum luminosity is known. This means that when such a supernova is seen in a distant galaxy the distance of the galaxy can be determined. Using observations of type 1a supernovae, distances out to three thousand million light years can be determined.

We have now established a foundation which enables us to understand how the age of the Universe has been determined.

6.3 Hubble and the Expanding Universe

An eminent worker in the field of measuring the distances of galaxies was the American astronomer Edwin Hubble (1889–1953), who both measured their distances by the methods previously described but also used the Doppler effect to measure their velocities along the line of sight. For some of the distant galaxies the velocities were a considerable fraction of the speed of light and for such velocities the Doppler shift

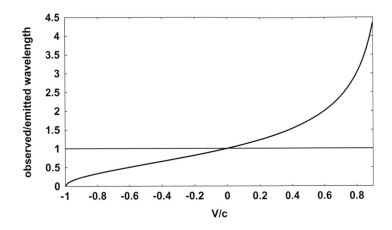

Figure 6.9　The relativistic Doppler effect for speeds comparable to the speed of light.

in wavelength is not given by the classical formula (6.6): a different equation must be used — the *relativistic Doppler-shift equation*, which is of the form

$$\frac{\lambda_o}{\lambda_s} = \sqrt{\frac{1 + V/c}{1 - V/c}},\qquad(6.7)$$

where λ_s is the wavelength emitted by the source and λ_o is that observed.

　　Figure 6.9 shows the change in wavelength as a function of the velocity expressed as a fraction of the speed of light. Negative velocities indicate motion towards the observer. For velocities of small magnitude that are much less than the speed of light the result from the relativistic equation is little different from that given by (6.6). However, the result departs appreciably at much higher velocities — for example, equation (6.6) indicates that for a body moving away from an observer at the speed of light the observed wavelength is just twice that it would be for a source at rest while the relativistic equation shows that the observed wavelength is infinite. During the early days of analysing spectral lines, workers were often puzzled by spectral lines they could not explain. This was because of large wavelength shifts due to large velocities, which the early workers were not expecting. For example, an emitted ultraviolet spectral line of wavelength 250 nm might be observed as a green line of wavelength 500 nm. Once they realized that they were dealing with very high velocities, they could then make sense of their observations.

Edwin Hubble

Hubble found that for galaxies not close to the Milky Way, which is a member of an association of nearby galaxies known as the *Local Group*, the galaxies he observed were *moving away from the Earth with speeds that were proportional to their distance*. This important result established the astronomical topic known as *cosmology*, the study of the Universe. From Hubble's results it follows that if a galaxy at a distance of three hundred million light years was receding at a speed of $6,000$ km s^{-1} then a galaxy twice as far away, at a distance of six hundred million light years, would be receding at $12,000$ km s^{-1}. The observation that speed of recession is proportional to distance is known as *Hubble's law*. Figure 6.10 shows the distances of galaxies against the speed of recession and it will be seen that the fit of points to a straight line is quite good. Hubble's law gives a way of determining the distances of galaxies that are so far away that even type 1a supernovae cannot be observed, or galaxies in which no type 1a supernova is occurring. If the law is valid at any distance, then the distances of these galaxies can be found by determining their speeds of recession by measuring Doppler shifts.

An inescapable conclusion from Hubble's observations is that *the Universe is expanding*. Again, if it is assumed that the motions of distant galaxies are along the line of sight from the Milky Way then, since

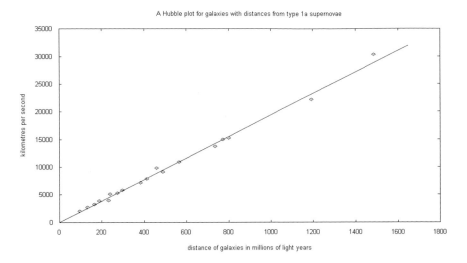

Figure 6.10   An illustration of Hubble's law.

distance is proportional to speed, if we imagine that time is reversed, at some time in the past all the matter in the Universe was concentrated in a small region of space. This has led to the *Big Bang Theory*, which considers that the Universe began with an enormous release of energy, initially concentrated at a point, which expanded, creating space and time and eventually all the mass that constitutes the tangible Universe.

The *Hubble Constant*, H_0 is the ratio of the speed of a galaxy V to its distance D, i.e.

$$H_0 = \frac{V}{D},\qquad(6.8)$$

and has the dimensions of inverse time. The reciprocal of the Hubble Constant is an estimate of the age of the Universe, based on it having begun at the time of the Big Bang. The latest age estimate from Hubble's constant is 14.4 billion years. However, this estimate has been modified by recent observations that suggest that the expansion of the Universe may be accelerating due to pressure forces exerted by a mysterious energy source, *dark energy*, which is thought to pervade the Universe. Taking this into account the current estimate of the age of the Universe is 13.8 billion years.

Chapter 7

The Ages of Globular Clusters and Young Stars

7.1 Early Material in the Universe

Following the Big Bang that created the Universe, there was a period when energy was being transformed into matter. We are accustomed to thinking of destroying matter to create energy, as happens in generating nuclear energy, but under some conditions the reverse can also occur. The first matter created by the Big Bang consisted of exotic particles like quarks but eventually the kind of matter that exists today was formed. This conventional matter was the nuclei of the simplest atoms — mainly hydrogen and helium, with some deuterium, an isotope (Section 8.2) of hydrogen, and perhaps some lithium. The composition of these nuclei, consisting of protons and neutrons, is shown in Figure 7.1.

Astronomers refer to this primaeval composition of the Universe as Population III material, somewhat illogically, because as we shall see it is followed successively by the formation of Population II and Population I material. Some of the Population III material formed into clouds of gas of somewhat higher density than the surrounding medium and then the clouds collapsed eventually to form stars. These early stars were probably quite massive and would have had a very short lifetime (Section 7.4), eventually exploding as supernovae and leaving remnants in the form of either neutron stars or black holes (Section 7.3). The energy and high temperatures generated in a supernova gave the conditions under which nuclear reactions could take place that generated heavier elements and these were injected into

113

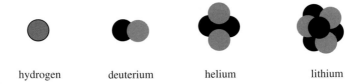

hydrogen   deuterium   helium   lithium

Figure 7.1   Light nuclei formed early in the expansion of the Universe (grey = proton, black = neutron).

          (a)   (b)

Figure 7.2   (a) The globular cluster M13. (b) The open cluster the Pleiades.

the Universe and became components of future stars. This new material composition — the original primaeval material plus a small addition of heavier elements — is referred to as Population II material.

7.2 Clusters of Stars

In Section 6.2.2 we mentioned the existence of field stars that move through the galaxy unaccompanied by any stellar companion. Also described were binary systems, either resolvable in a telescope and called visual binaries or unresolvable, referred to as spectroscopic binaries. However, many stars exist as members of associations of many stars in *stellar clusters*, which are of two types. The first type is the *globular cluster*, containing several hundred thousand stars; one such cluster, named M13, is shown in Figure 7.2a. Stars are unresolved in the heart of the cluster but individual stars can be seen in the outer regions. The other type, consisting of anything from a hundred to a few thousand stars, are known as *open clusters* or, sometimes, *galactic clusters*. A striking example, shown in Figure 7.2b, is the *Pleiades*, consisting of about 500 stars of which seven are very bright. It was well known in antiquity and there are biblical references to it. An alternative name from Greek mythology is *Seven Sisters*, relating to the bright stars.

The material between stars and clusters is known as the *interstellar medium* (ISM) — a crude description of which is that it is nearly nothing — but not quite. In each cubic centimetre of the ISM there is one atom, mostly either of hydrogen or helium. It also contains dust particles, typically one micron in diameter and, on average, there is one dust particle per cubic kilometre of the ISM. It seems so little — yet it is an important part of many astronomical bodies, including Earth and all in it, inanimate and living.

There is a sense in which field stars are members of a stellar association — the galaxy in which the stars move. In the case of the Sun the galaxy is the Milky Way, a collection of one hundred thousand million stars, well separated from other galaxies, of which there are an estimated one hundred thousand million in the Universe.

By analysing the spectral composition of the light coming from stars it is possible to find their composition in terms of the atomic species they contain. Astronomers have looked in vain for Population III stars that contain just the primordial material from the formation of the Universe. The stars in globular clusters are found to contain some, but very little, heavier material, usually much less than 0.2% by mass, and the assumption is that this material was ejected from the Population III star supernovae and then mixed with the primordial gas. Stars with this kind of composition are known as Population II stars. By contrast, the stars in open clusters, known as Population I stars, which are similar to the Sun, contain up to 2% of heavier material that must have come from several cycles of star formation and destruction.

The formation of associations as large as galaxies by a single mechanism has presented astronomers with insuperable problems. This has to do with a concept known as the *Jeans critical mass*, first introduced by the British astronomer James Jeans (1877–1946), which gives the *minimum* mass M_J of gaseous material at density ρ and temperature T that can collapse to form a condensed body. It is of

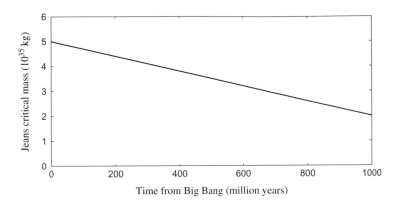

Figure 7.3   The Jeans critical mass for Universe material as a function of the age of the Universe.

the general form

$$M_J = C \sqrt{\frac{T^3}{\rho}}, \tag{7.1}$$

where C is a constant that depends on the material of the gas. However, although the formula gives the minimum mass that *can* condense it is also a good approximation to the mass that would *actually* condense due to instability in a large expanse of material. There are theoretical models of the way that density and temperature would change as the Universe expands and hence it is possible to calculate M_J as a function of time from the Big Bang. This is shown in Figure 7.3; the Jeans critical mass varies very little over the first 1,000 million years, falling from about 5×10^{35} kg to about 2×10^{35} kg. The mass of the Sun is approximately 2×10^{30} kg and an average main-sequence star has a mass 1.4×10^{30} kg, so the masses of material that would condense in the early universe would have the mass of a few hundred thousand stars — typical of a globular cluster.

The current belief is that globular clusters were first formed and that large numbers of them clumped together to form galaxies. This model suggests that the formation of globular clusters was the next significant event in the evolution of the Universe after the demise of the first Population III stars, so there is considerable interest in determining the age of globular clusters. However, to do this we need to know about the processes that occur in the evolution of stars.

7.3 The Birth, Life and Death of a Star

The processes that operate in forming a star and in its subsequent development are little affected by a small amount of heavier material so the same description applies to Population III, II and I stars. We consider a time when a galaxy has formed by the amalgamation of globular clusters and there is an ISM occupying all the space within the galaxy not occupied by stellar clusters or individual stars. A supernova explosion of a star will inject dusty material into the surrounding ISM and also compress it, which will have the effect of enabling various cooling processes to occur that reduce the local temperature.[1] Cooling a gas lowers its pressure so the cooled region is further compressed by the effect of the larger pressure of the surrounding gas at a higher temperature. Compressing the cooled region increases its density and increasing density leads to enhanced cooling. Eventually, the compression of the region stops because, although higher density leads to more cooling, lowering the temperature leads to less cooling and eventually the effect of decreasing temperature balances that of increasing density. What we then have is a dense cool cloud (DCC) within the ISM (Figure 7.4). It is within these clouds that star formation takes place. It must be stressed that in an astronomical context the word 'dense' must be understood relative to the density of the ISM. The mean density of the ISM is about 10^{-21} kg m^{-3} and a dense cloud might have one thousand times that density. To put this in perspective, some physicists carry out experiments within what they call an 'ultra-high vacuum', which is a million times denser than the dense clouds within which stars are formed.

When the dense clouds form, they begin to collapse under self-gravitational forces. The mode of collapse is called *free-fall*; it starts very slowly, almost imperceptibly, but it accelerates and in the final stages it is extremely rapid. Although compressed gas usually heats up (like the air in the barrel of a bicycle pump when a tyre is being inflated), there are cooling processes that radiate energy out of the cloud and keep it

---

[1] Information about cooling processes can be found in the author's book *The Formation of the Solar System: Theories Old and New*, 2007, pp. 164–169 (London: Imperial College Press).

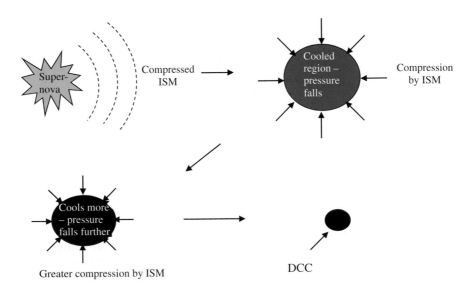

Figure 7.4 The stages in the formation of a DCC, triggered by a supernova.

cool. When the collapse of the cloud becomes sufficiently rapid, the motion of the gas becomes turbulent so that within the general inward motion of the gas there are streams of gas moving haphazardly in various directions. This kind of phenomenon, in which a quickly moving fluid becomes turbulent, can be seen when a slowly flowing river in a wide channel enters a narrow ravine. The water flows more quickly and turbulence sets in, producing 'white water' — conditions that provide an exciting environment for canoeing or rafting for some adventurous individuals. In the gas clouds, turbulent streams occasionally collide head-on, and when they do so, they further compress the gas in the collision region. This compressed region can, under favourable circumstances, become an embryonic star, called a *protostar*, which is in the form of a ball of gas of total mass from a fraction to several solar masses, with a radius of order 2000 au, 60 times the radius of the Solar System, and with a temperature in the range 20–50 K.

7.3.1 *Collapse to the main sequence*

The collapse of a protostar of solar mass to the main sequence was described by the Japanese astrophysicist Chushiro Hayashi

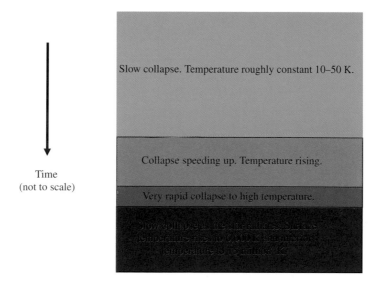

Figure 7.5 Stages in the collapse of a protostar.

(1920–2010). We can follow the stages in the collapse in Figure 7.5. The free-fall collapse time for a protostar of solar mass and radius 2,000 au is about 20,000 years. The collapse is initially at a constant low temperature because the heat generated can easily be radiated out of the almost transparent protostar. This stage lasts for most of the free-fall time and is the blue stage in Figure 7.5.

As the collapse proceeds, the protostar material becomes more opaque to radiation, trapping more of the generated heat so that the temperature increases; this is the orange stage in Figure 7.5, with the final temperature about 100 K. Now the collapse becomes rapid, the opacity of the material greatly increases and the temperature rapidly rises. The surface temperature increases to several thousand K and the interior temperature becomes much higher than that. Because of both the high density and high temperature the interior pressure builds up, opposing the force of gravity, until eventually the pressure and gravity forces are in balance and the collapse ceases. The protostar has developed into what is called a *young stellar object* (YSO).

Although we have stated that the collapse ceases, this is not really so. What actually happens is that the collapse becomes very much slower and is mediated by a different mechanism. The YSO radiates energy and its reaction to this is to collapse slowly and, paradoxically, to become *hotter*. The gravitational energy released by the collapse is sufficient to provide both energy emitted as electromagnetic radiation and energy to heat up the YSO. For a YSO with the mass of the Sun, the collapse causes the surface temperature to increase from about 3,000 K to 6,000 K and the interior temperature to rise to 15 million degrees. At this temperature, nuclear reactions take place converting hydrogen into helium and generating large amounts of heat in the process. The star has now reached the main sequence stage — the end of the red stage in Figure 7.5. With energy being generated in its core, the star can provide the radiated energy without the need to collapse further — in the case of a solar-mass star for about 10,000 million years.

7.3.2 *The final journey to obscurity*

The highest temperature in the star is at its centre so that is where nuclear reactions are most rapid and where hydrogen eventually becomes depleted. Since the temperature is too low for nuclear reactions involving helium to take place, the pressure in the core falls and the core begins to contract under the pressure of the overlying material. This contraction releases gravitational energy which heats up the region round the centre thus enabling nuclear reactions to occur in a still hydrogen-rich shell of material. This shell gradually expands as the hydrogen in successive regions becomes exhausted; the star has entered the stage of *hydrogen-shell burning*, illustrated in Figure 7.6.

The pressure exerted by the shell burning pushes both outwards and inwards The outward push causes expansion of the star and reduction of its surface temperature — the star has become a *red giant* with radius about one astronomical unit. The inward push compresses the core, which at this stage is virtually all helium, and the temperature of the core increases. Eventually the temperature in the core reaches one hundred million kelvin and nuclear reactions in which three helium nuclei are transformed into a carbon nucleus begin to take place — called the *triple-alpha reaction* because a helium nucleus is an alpha-particle (Figure 7.7). This new and powerful source of energy heats up the core giving an increasing reaction rate.

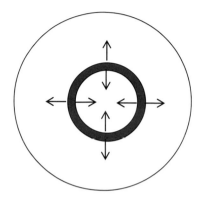

Figure 7.6 Hydrogen-shell burning: Pressure forces, acting in the directions shown by the arrows, compress the core (together with gravity) and expand the outer parts of the star to produce a red giant.

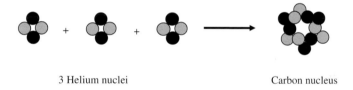

3 Helium nuclei   Carbon nucleus

Figure 7.7 The net result of the triple-alpha reaction (grey = proton, black = neutron).

With helium reactions and energy generation in the core, the situation is similar to that in the main-sequence stage and the star shrinks to a configuration somewhat like that of the main sequence. Eventually the helium in the core is depleted and helium-shell burning sets in, again compressing the core and expanding the star. During the final stages of helium-shell burning the outward pressure ejects outer star material in the form of shells which, illuminated by the star, look like rings (Figure 6.8). Such objects are known as *planetary nebulae,* although they have nothing to do with planets.

The development of the star after helium-shell burning depends on the star's mass. For a star of about solar mass the outer layers of the star are completely stripped off, leaving the compressed core in the form of a *white dwarf.* The material of a white dwarf, mostly carbon, is very unlike normal material. It is in a high-density *degenerate state*; a state of matter where large gravitational forces are balanced by forces due to quantum-mechanical effects. A white dwarf has the mass of the Sun in a body the size of the Earth. No nuclear reactions take place in it; it slowly cools, eventually to become a non-luminous *black dwarf.*

Stars of greater mass pass through the main-sequence stage more quickly and do not end up as white dwarfs. Temperatures in the core become much greater and further nuclear reactions take place, producing ever heavier nuclei up to the nucleus of iron. The star goes through a series of expansions and contractions, but eventually the pressure in the core becomes so great that protons and electrons are squeezed together to become neutrons. At that stage, the core, which now consists only of neutrons, rapidly collapses to give a small core of tightly packed neutrons and outer material streams in to occupy the empty space. It bounces off the neutron core and crashes into material that is still moving inwards. The result is a supernova explosion that expels all the material outside the neutron core. The debris from a supernova, the Crab Nebula first observed in 1054, is shown in Figure 7.8. The dense residue becomes a *neutron star*; with twice the mass of the Sun it would have a diameter of a few kilometres.

For stars with masses above about six solar masses the pressure in the core becomes so great that even neutrons become compressed and the result is then a *black hole,* an object with finite mass and zero

Figure 7.8   The Crab Nebula.

dimension (at least in principle) which is invisible because light cannot escape from its gravitational field.

7.4 The Duration of the Main Sequence and the Age of Globular Clusters

It might seem logical that more massive stars, with more hydrogen available as fuel, would spend longer on the main sequence, but this is not so. The more massive the star, the higher is the temperature in its core and the faster is the rate of nuclear reactions. As a first approximation, the rate at which hydrogen is converted to helium varies as $M^{7/2}$, where M is the mass of the star, and the net result is that the lifetime of the main-sequence stage varies as $M^{-5/2}$; a twofold increase in mass decreases the lifetime by a factor of six.

Of course, we cannot directly measure main-sequence lifetimes, but there have been detailed theoretical calculations, the results of which are shown in Table 7.1. The times for the stages of star formation, from first forming a protostar through the lifetime as a YSO, are small compared with the main-sequence lifetimes, so the main-sequence lifetimes in Table 7.1 are effectively the time from star formation to the end of the main sequence.

Table 7.1 Lifetime on the main sequence as a function of stellar mass.

Stellar mass (solar units)	Main-sequence lifetime (10^6 years)	YSO lifetime (10^6 years)
15	10	0.062
9	25	0.15
5	100	0.58
3	350	2.5
2.25	900	5.9
1.5	2,000	18
1.25	4,000	29
1	10,000	50
0.5	50,000	150

The fact that a star is on the main sequence can be determined by looking at its spectrum, which not only gives its temperature but all other physical characteristics, including its mass (Section 6.2.2). When the stars in globular clusters are examined it is found that only main-sequence stars with a mass less than 0.8 solar masses are observed. The implication of this observation is that all stars of greater mass have passed through the main-sequence stage and have become white dwarfs, neutron stars or black holes. Of course, it is difficult to *precisely* determine the cut-off mass for the observation of main-sequence stars in globular clusters and there are also some uncertainties in the main-sequence lifetimes as deduced from theory. However, most estimates of the ages of globular clusters are around 13.5 billion years, not much less than the 13.8 billion years given in Section 6.3 as the age of the Universe.

The picture that emerges is that soon after the Big Bang massive Population III stars formed, moved through their main-sequence lifetimes in less than 200 million years and the resulting heavier elements produced were sufficient to give the low heavy-element composition of the Population II stars that populate globular clusters. Large numbers of globular clusters then amalgamated to give galaxies of various types and sizes. Later in the evolution of the Universe, after several generations of higher-mass stars producing heavier elements, open clusters were able to form with Population I stars containing up to two percent of heavier elements.

7.5 The Age of Evolving Young Stellar Objects

Once a star reaches the main sequence, its external appearance changes very little until it embarks on its final journey towards being a white dwarf, a neutron star or a black hole. For that reason, it is not possible to determine the age of a main-sequence star; to determine age there must be some process of change to observe. However, changes of properties do occur in the development of a star from the beginning of its YSO stage when it starts to radiate energy, causing it to collapse and increase in temperature until hydrogen-based reactions begin, at which stage it enters its main-sequence stage.

To understand how the age of a YSO may be determined, we first need to describe an important tool in astronomy, the *Hertzsprung–Russell* (*H–R*) *diagram*, independently devised by the Danish astronomer Ejnar Hertzsprung (1873–1967) and the American astronomer Henry Norris Russell (1877–1957). In Section 6.2.2, the way in which the temperature and luminosity (total energy output per unit time, or brightness) of a star can be found was described, and the Hertzsprung–Russell diagram is a plot of the temperature of a star on the x-axis against the luminosity on the y-axis. When a large number of stars are plotted in this way, say those closest to the Sun, it is found that stars do not appear uniformly over the plot but occupy distinct regions. Figure 7.9 is a schematic of an H–R diagram showing these regions.

The luminosity of a star increases with both temperature and its radius so that the position on the H–R diagram indicates both the temperature and size of the star. The position of the Sun in the main sequence is indicated by the orange star. At the same temperature there are stars with more than 10,000 times the luminosity, which indicates that they must be very large — *supergiants*. There are also stars that have

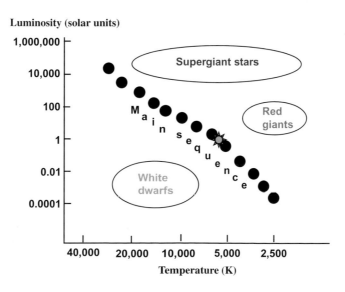

Figure 7.9 A schematic of a Hertzsprung–Russell diagram.

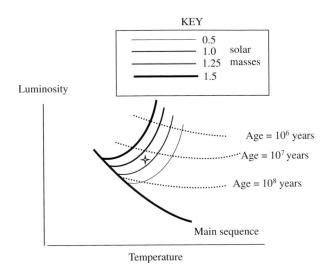

Figure 7.10   Pathways of YSOs towards the main sequence.

a much higher temperature than the Sun but are much less luminous; these are white dwarfs.

The progress of a star towards the main sequence, while it is a YSO, can be theoretically deduced. The temperature and luminosity of the YSO depends on both its mass and the time when it first became a YSO — which defines its age. These theoretical paths are shown in Figure 7.10. For the four star masses the pathway to the main sequence is shown together with contours that indicate the time it has taken the YSO to reach different points. The measured temperature and luminosity of a YSO determines both its mass and its age; the observation represented by the cross in Figure 7.10 shows that the YSO has a mass of about 1.1 solar masses and an age of approximately thirty million years.

Over the years, there have been different theoretical models for the evolution of YSOs, but while they differ in detail they all broadly give the same pattern of development. The final column of Table 7.1 gives YSO lifetimes and it will be seen that massive stars spend little time as YSOs as well as comparatively short times on the main sequence.

One must be cautious about interpreting the state of a star away from the main sequence. In the stage where a star is moving towards being a red giant, after the main-sequence lifetime has been completed, a star moves up and towards the right of the H–R diagram in the same region as that occupied by YSOs moving *towards* the main sequence. However, YSOs are normally part of a forming open cluster and if the cluster contains large mass stars on the main sequence, then, from Table 7.1, we may reasonably deduce that the cluster is young and that what we are observing are YSOs.

Chapter 8

The Age of the Solar System

8.1 A Brief Description of the Solar System

The dominant body of the Solar System is the Sun, a normal Population I main-sequence star that contains 99.86% of the total mass of the system. Stars change little in their externally observed characteristics during the main-sequence stage so it is impossible to determine their ages by measurements of temperature and luminosity. For that reason, in assessing the age of the Solar System what is done is to estimate the age of some of the material that orbits the Sun and to take the oldest estimate as the age of the whole system, the Sun included. The assumption made here is that the planets and other bodies orbiting the Sun are coeval, or nearly so, with the Sun, an assumption that is consistent with most theories of the origin of the Solar System.

Next in size and mass are the planets, which form two main groups, the *terrestrial planets* closest to the Sun — Mercury, Venus, Earth and Mars — and the *major planets* further out — Jupiter, Saturn, Uranus and Neptune. All the planets other than Mercury and Venus have orbiting satellites, of which Earth's Moon is the most familiar. The sizes and mean orbital radii of the planets are shown in Figure 8.1.

Other largish bodies in independent heliocentric orbits are the *dwarf planets* — Ceres, Pluto, Eris, Haumea and Makemake — with sizes characteristic of some of the larger, but not the largest, satellites. Ceres orbits in the region between the terrestrial and major planets and Pluto mostly orbits outside Neptune but at its closest to the Sun comes just inside Neptune's orbit. The other three orbit outside Neptune in a region known as the *Kuiper Belt*.

129

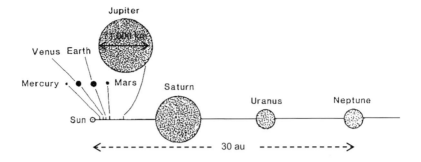

Figure 8.1 The orbital radii and sizes of the planets (different scales).

The Solar System contains a great deal of smaller-scale solid material in the form of asteroids, consisting of silicate, iron or a mixture of the two, and comets, consisting of loose aggregations of silicate and volatile materials such as water-ice, solid methane, ammonia and carbon dioxide. There is a large reservoir of comets in the Kuiper-belt region that occasionally are perturbed by Neptune into the inner Solar System. When they approach the Sun, the icy material vaporizes and the comet develops a tail pointing away from the Sun. There is another vast reservoir of comets — the *Oort Cloud* — much further out at distances of a few tens of thousands of au.

The asteroids, varying in dimension from a few metres to a few hundreds of kilometres, are mostly in the *asteroid belt*, the region between the terrestrial and major planets, within which Ceres orbits. Occasionally, there are collisions between asteroids and chips fly off into space. These fragments of asteroids, called *meteorites*, land on Earth in large numbers and, when they can be recognized for what they are, provide a free but valuable sample of the material orbiting the Sun. They also contain information regarding their ages, i.e. how long ago they first consolidated into a rock-like form, and the oldest ages found for meteorites are taken as indicators of the age of the Solar System.

8.2 Radioactivity

The key to determining the ages of meteorites is that they contain radioactive *isotopes* of some elements. To understand what an isotope is

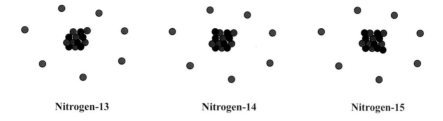

Nitrogen-13   Nitrogen-14   Nitrogen-15

Figure 8.2   The structures of nitrogen-13, nitrogen-14 and nitrogen-15.

we show a representation of three isotopes of nitrogen in Figure 8.2 — nitrogen-13, nitrogen-14 and nitrogen-15. In each case the central nucleus contains seven positively charged protons (red); it is the number of protons that determines what kind of atom it is — in this case nitrogen, which has the *atomic number* 7. Also present in the nucleus are neutrons (black), with no charge and a mass about the same as that of the proton. The number of neutrons is six for nitrogen-13, seven for nitrogen-14 and eight for nitrogen-15. Protons and neutrons are collectively called *nucleons* and the total number of nucleons gives the *atomic mass* — 13, 14 and 15. Finally, to complete an electrically neutral atom there are seven electrons (blue) with a negative charge equal in magnitude to that of the proton, so giving a neutral atom with no net charge. The nucleus is very small, typically a few times 10^{-15} m (a thousand-million-millionth of a metre) across, while the electrons occupy a region of size a few times 10^{-10} m (a ten-thousand-millionth of a metre) across. To put this in perspective, if a nucleus were expanded to the size of a tennis ball, the average distance of an electron from the nucleus would be about five kilometres.

Most living systems contain nitrogen and while most of that nitrogen is nitrogen-14, 0.4% of it is nitrogen-15. Both nitrogen-14 and nitrogen-15 are stable, which means that they will never change into anything else. However, nitrogen-13 is unstable — a radioactive isotope — and the form of *decay* it exhibits is that one of the protons in the nucleus transforms into a neutron plus a *positron*, a particle with the same mass as an electron but with a positive charge. The positron escapes from the nucleus with high speed, leaving behind a nucleus, the *daughter product*, with six protons and seven neutrons, which is an isotope of carbon, carbon-13. In order to be neutral carbon, it also has to shed one of its electrons.

The nitrogen-13 atoms do not all convert into carbon-13 immediately or simultaneously. They do so with a *half-life* of ten minutes, meaning that at the end of a ten-minute period the number of nitrogen-13 atoms remaining is one half the number at the beginning of the period. So, if we began with 2,000 nitrogen-13 atoms, then after 10 minutes there would be 1,000, after 20 minutes 500, after 30 minutes 250 and so on. For this reason nitrogen-13 does not occur in nature, but it can be formed in a nuclear reactor. The half-lives of isotopes can vary from fractions of a second to billions of years and it is isotopes with longer half-lives that enable us to determine the ages of meteorites and hence of the Solar System.

8.3 Radioactive Dating: A Simple Procedure

Let us consider a mineral containing a radioactive isotope A (*parent*) that decays to give a *daughter product* B. The way that the quantities of A and B vary with time, measured in units of the half-life, is shown in Figure 8.3. The residual amount of A is given by the distance below the line and that of B by the distance above the line; it is clear that the ratio of the amount of A present to that of B gives a measure of the

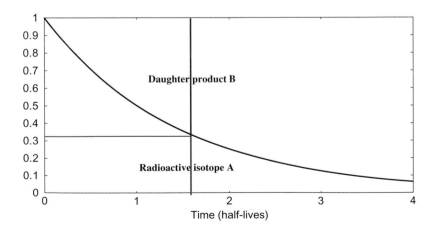

Figure 8.3 The decay of a radioactive isotope.

time from when the mineral became a *closed system*, i.e. it neither lost any material nor gained it from an external source. As an example, the vertical line in Figure 8.3 shows the amounts of A (blue) and B (red) present at some time and gives $A/B = 1/3$. Translated into an age it is 1.585 half-lives which, given the half-life, can be expressed in some unit of time.

This is the simplest of all methods for measuring the age of rocks but its validity depends on two factors. The first is that the daughter product should be retained in the mineral; if it is a gas, then some may be retained in interstices within the body of the mineral but some may be lost, leading to an age estimate that is too low. The second condition is that daughter product B should not be present in the newly formed mineral, because this would enhance the amount of B present and lead to an overestimate of the age. There is a temperature above which trapped gases cannot be retained by the mineral. This is the *closure temperature*. Some values for various minerals are collected in Table 8.1. Age determinations using trapped gases produced by radioactivity in the rock then refer back to the time the rock cooled to this temperature. It is seen that these temperatures vary over a wide range: it is 750°C in the mineral zircon for uranium decaying to lead measurements (Section 8.6.3) down to 280°C for biotite using the potassium-argon method. The most accurate radioactive decay scheme is where the half-life is comparable to the age being measured.

Table 8.1   Some closure temperatures for three dating methods.

Parent to daughter	Mineral	Closure Temperature (°C)
Potassium to argon	hornblende	540
	biotite	280 ± 40
	muscovite	≈ 350
Uranium to lead	zircon	>750
	monazite	>650
	apatite	≈ 350
Rubidium to strontium	biotite	320
	muscovite	>500
	feldspar	≈ 350

8.3.1 *Reaction chains*

So far we have dealt with the simplest case when there is a single daughter, but for heavier radioactive elements the daughters themselves have daughters. Then it is necessary to follow the decay chain an appropriate number of times until the stable end product is reached. If the half-lives of all the decays in the chain are known, then it is possible to estimate an age for the mineral, but it is a complicated process requiring computer calculations. There is one situation, however, where the solution is simple. That is when the half-life of one of the decay steps is substantially larger than that of any of the others in the chain. This is the case, for example, for the decay of uranium-238 which goes through 19 decay steps before the final daughter product — lead-206 — is formed. The first decay step gives thorium-234 with a half-life of 4.51 billion (10^9) years, or about the present age of the Solar System. This is substantially greater than the half-life of any other decay in the chain, the greatest of which is 245,500 years. Under this circumstance it may be assumed that there is a single decay from parent uranium-238 to daughter lead-206 with half-life 4.51 billion years.

8.4 Some Important Decays for Age Determination

There are many radioactive decays that can be used for age determination, some of which are listed in Table 8.2. In Section 8.6 we shall just describe two of them and the way they are used.

For the greatest accuracy in determining the age of a mineral, be it from Earth or contained in a meteorite, it is desirable if possible to use several methods and to take an average of the ages found.

8.5 Rocks, Grains and Minerals

An individual rock normally contains a variety of minerals which have crystallized from the same original molten material. The *whole-rock content* of rocks will normally be categorized by their chemical composition, which is done by breaking it down into components, mainly oxide components. For example, the chemical formula for the

Table 8.2 Parent/daughter decays of geologically interesting radioactive isotopes.

Parent	Daughter	Half-life (million years)
uranium-238	lead-206	4.468
uranium-235	lead-207	704
thorium-232	lead-208	14,010
rubidium-87	strontium-87	48,800
samarium-147	neodymium-143	106,000
potassium-40	calcium-40	1,250 combined
	argon-40	
argon-39	potassium-39	0.000269
lutetium-176	hafnium-176	37,300
rhenium-187	osmium-187	42,000
carbon-14	nitrogen-14	0.00573

very common mineral *olivine* is $(Mg,Fe)_2SiO_4 = 2(Mg,Fe)O + SiO_2$ (Mg = magnesium, Fe = iron, Si = silicon, O = oxygen). The bracketed part of the formula indicates that in the mineral iron and magnesium are interchangeable. An alternative way of expressing the composition of an olivine crystal grain is $Mg_xFe_{2-x}SiO_4$, where x can vary between 0 and 2 for any particular crystal grain. Olivine is a member of a class of *ferromagnesian minerals* with a large component of iron or magnesium plus SiO_2 but with other components sometimes present. Other ferromagnesian minerals are: *augite*, $(Ca,Na)(Mg,Fe,Al,Ti)(SiAl)_2O_6$ (Ca = calcium, Na = sodium, Al = aluminium, Ti = titanium), and *diopside*, $CaMgSi_2O_6$.

Rocks consist of a myriad of grains of different minerals compressed together; Figure 8.4 shows the appearance of the common rock granite under a microscope. There are three main types of rock, namely:

Igneous rocks

These are formed by the cooling of molten material (*magma*) and the grain sizes depend on the cooling rate. Slow cooling enables grains to grow to a large size, giving a coarse structure, whereas rapid cooling produces grains so small that to the eye the material seems smooth and homogeneous.

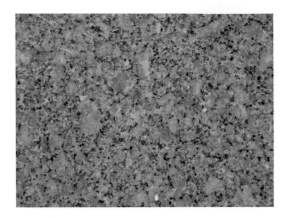

Figure 8.4   The granular structure of a sample of rose granite.

Sedimentary rocks

As their name suggests, these are formed by accumulated sediments deposited by the settling of fine grains carried by water or transported by wind.

Metamorphic rocks

These were originally either igneous or sedimentary rocks that have been subjected to high temperature or high pressure or both for long periods. They often show swirl-like patterns that make them easily recognizable.

It is important to understand that a given rock will contain grains of different minerals and also that mineral grains of the same type, i.e. olivine or augite, may have a range of chemical compositions.

8.6 Examples of Dating Using Radioactive-Decay Systems

It is clear from Table 8.2 that there are many radioactive decays that can be used for dating rocks. Many of them depend on the presence of isotopes of atoms that are not particularly rare in nature but are not well known, such as rubidium, which is actually more common than copper, with which we are very familiar. Here we describe two radioactive systems that are often used for dating rocks, either from Earth or coming from space.

8.6.1 *The rubidium → strontium system*

The half-life of rubidium is 48.8 billion years so this decay is best used for old rocks. Rubidium (chemical symbol Rb) has two isotopes of masses 85 (stable) and 87 (radioactive) while strontium (chemical symbol Sr) has four isotopes of masses 84, 86, 87 and 88. The decay process of interest is that of Rb-87 to Sr-87. The simple method illustrated in Figure 8.3 would involve measuring the amounts of the parent Rb-87 and of the daughter product Sr-87 and using the ratio to determine the age. It should be mentioned at this point that age determinations are actually carried out mathematically, using the amounts of parent and daughter, together with the half-life to determine the age. Here we shall describe the process by which the age is determined without the mathematical details that explain the basis of the process. In the next section the mathematical basis of the method will be described.

Strontium occurs fairly widely in rocks and it is quite likely that there was some initial strontium present when the rock first formed, although it is not known how much. At the present time the ratios of the strontium isotopes are:

$$Sr\text{-}84\ (0.56\%) \quad Sr\text{-}86\ (9.86\%) \quad Sr\text{-}87\ (7.00\%)$$
$$Sr\text{-}88\ (82.58\%).$$

Since Sr-86 is not a product of any radioactive decay the amount at the present time should the same as that in the mineral sample initially and this gives the key for age determination. The quantities that can be measured (Section 8.8) are the present concentration of Rb-87 in the rock, $[Rb\text{-}87]_{now}$,[1] the present concentration of Sr-87 in the rock $[Sr\text{-}87]_{now}$ and the concentration of Sr-86, $[Sr\text{-}86]_{now}$, which is the same as the original concentration, $[Sr\text{-}86]_0$. Plotting the ratio of $[Sr\text{-}87]_{now}/[Sr\text{-}86]_{now}$ against the ratio $[Rb\text{-}87]_{now}/[Sr\text{-}86]_{now}$ gives

---

[1] Concentration is the number of atoms per unit volume.

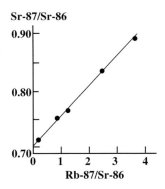

Figure 8.5   A whole rock isochron from an old rock in Greenland.

a straight line, the slope of which, in conjunction with the known half-life, gives the age. Such a plot, known as a *whole rock isochron*, is shown in Figure 8.5. The best results are obtained when the samples being measured have a wide range of rubidium concentrations. Because of inevitable errors of measurement the points do not fall exactly on a straight line but the best straight line can be calculated mathematically. The intercept of the straight line on the *y*-axis gives the *initial ratio* of $[\text{Sr-}87]_0/[|\text{Sr-}86]_0$, when the rock formed, and the deduced age of the ancient Greenland rock, from the slope of the line and the known half-life, is 3.66 billion years.

8.6.2 *A mathematical treatment of the rubidium → strontium system*

As an alternative to describing the rate of decay in terms of the half-life, there is another quantity, the *decay constant*, λ, such that the present concentration of rubidium after time t $[\text{Rb-}87]_t$ is related to the original concentration $[\text{Rb-}87]_0$ by

$$[\text{Rb-}87]_t = [\text{Rb-}87]_0 \exp(-\lambda t) \qquad (8.1a)$$

or, alternatively,

$$[\text{Rb-}87]_0 = [\text{Rb-}87]_t \exp(\lambda t). \qquad (8.1b)$$

The relationship between the decay constant and the half-life, $\tau_{1/2}$, is found from

$$\exp(\lambda\tau_{1/2}) = \frac{[\text{Rb-87}_0]}{[\text{Rb-87}]_{\tau_{1/2}}} = 2$$

or

$$\lambda\tau_{1/2} = \ln(2) = 0.6931. \qquad (8.2)$$

The amount of strontium produced by the decay over the lifetime of the rock is

$$[\text{Sr-87}]_{\text{decay}} = [\text{Rb-87}]_0 - [\text{Rb-87}]_{\text{now}}$$

and using (8.1b) this becomes

$$[\text{Sr-87}]_{\text{decay}} = [\text{Rb-87}]_{\text{now}}\{\exp(\lambda t_a) - 1\}, \qquad (8.3)$$

where t_a is the age of the rock.

The total amount of strontium present in the rock is the sum of that produced by decay and that originally present and is

$$[\text{Sr-87}]_{\text{now}} = [\text{Rb-87}]_{\text{now}}\{\exp(\lambda t_a) - 1\} + [\text{Sr-87}]_0. \qquad (8.4)$$

Equation (8.4) contains two unknown quantities — t_a and $[\text{Sr-87}]_0$ — the latter of which varies from one mineral grain to another depending on the strontium content, so the age cannot be found from this equation. We now divide the equation throughout by the initial amount of Sr-86, $[\text{Sr-86}]_0$, which is called a *normalizer* and since $[\text{Sr-86}]_{\text{now}} = [\text{Sr-86}]_0$, it gives

$$\frac{[\text{Sr-87}]_{\text{now}}}{[\text{Sr-86}]_{now}} = \{\exp(\lambda t_a) - 1\}\frac{[\text{Rb-87}]_{\text{now}}}{[\text{Sr-86}]_{\text{now}}} + \frac{[\text{Sr-87}]_0}{[\text{Sr-86}]_0}. \qquad (8.5)$$

The final term is now constant from one mineral grain to another since no matter how much strontium was present the ratio $[\text{Sr-87}]_0/[\text{Sr-86}]_0$ was the same. We now consider (8.5) as a linear equation of the form $y = mx + c$, where

$$y = \frac{[\text{Sr-87}]_{\text{now}}}{[\text{Sr-86}]_{\text{now}}}, \quad m = \{\exp(\lambda t_a) - 1\},$$

$$x = \frac{[\text{Rb-87}]_{now}}{[\text{Sr-86}]_{\text{now}}} \quad \text{and} \quad c = \frac{[\text{Sr-87}]_0}{[\text{Sr-86}]_0}.$$

Both x and y can be found for each mineral sample in the rock and when all the values are plotted, they will fall on a straight line of slope $m = \{\exp(\lambda t_a) - 1\}$ with the intercept on the y-axis giving the initial ratio, $[\text{Sr-87}]_0/[\text{Sr-86}]_0$. For the Greenland rock the initial ratio was found to be 0.701, slightly different from the current value.

One difficulty with the use of rubidium is that it is quite volatile and will escape from rocks if they are heated over long periods.

8.6.3 *Uranium → lead and thorium → lead*

This uses the decay chains uranium-238 to lead-206 and uranium-235 to lead-207. Lead (chemical symbol Pb) has four naturally occurring isotopes, two of which are formally radioactive but with such long half-lives that they are effectively stable. These isotopes are, with half-lives and percentage occurrences in parentheses:

Pb-204 (1.4×10^{17} years, 1.4%)   Pb-206 (stable, 24.1%)
Pb-207 (stable, 22.1%)   Pb-208 (2×10^{19} years, 52.4%).

The isotope lead-204 is not the product of any radioactive process and so can be used as a normalizer for producing an isochron plot for both of the uranium (chemical symbol U) to lead decays, as shown in Figure 8.5, in which [Pb-206]/[Pb-204] is plotted against [U-238]/[Pb-204]. However, another way of using these reactions is to measure Pb-204, Pb-206 and Pb-207 and to plot [Pb-207]/[Pb-204] against [Pb-206]/[Pb-204], a lead–lead isochron. Once again, the plot is linear and the slope of the best straight line is an indicator of the age of the rock. The advantage of this method is that only the concentrations of lead isotopes are measured and if there is any loss of lead due to heating of the rock it will be proportionately the same for all of them, leaving the ratios unaffected. The mathematics of this method is described in the following section.

There are some minerals, such as zircon ($ZrSO_4$) and titanite ($CaTiSO_5$), which in their formation cannot possibly accommodate lead in their crystal lattices and so initially contain no lead; in that case the simple procedure described in Section 8.3 can be used to estimate an age.

Another useful decay is that of thorium-232 (chemical symbol Th) to lead-208. Thorium and lead are less easily lost from minerals than uranium, making this an attractive dating procedure. The very long half-life for thorium means, however, that the method is not very useful for young rocks.

8.6.4 *The mathematics of the lead–lead isochron*

We start with the two equations

$$\frac{[\text{Pb-206}]_{\text{now}}}{[\text{Pb-204}]_{\text{now}}} = \frac{[\text{U-238}]_{\text{now}}}{[\text{Pb-204}]_{\text{now}}}\{\exp(\lambda_{238}t_a) - 1\} + \frac{[\text{Pb-206}]_0}{[\text{Pb-204}]_0}$$

$$(8.6a)$$

and

$$\frac{[\text{Pb-207}]_{\text{now}}}{[\text{Pb-204}]_{\text{now}}} = \frac{[\text{U-235}]_{\text{now}}}{[\text{Pb-204}]_{\text{now}}}\{\exp(\lambda_{235}t_a) - 1\} + \frac{[\text{Pb-207}]_0}{[\text{Pb-204}]_0}$$

$$(8.6b)$$

with t_a the age of the mineral. These equations can be combined to give

$$\frac{[\text{Pb-207}]_{now}/[\text{Pb-204}]_{\text{now}} - [\text{Pb-207}]_0/[\text{Pb-204}]_0}{\|\text{Pb-206}\|_{\text{now}}/[\text{Pb-204}]_{\text{now}} - [\text{Pb-206}]_0/[\text{Pb-204}]_0}$$

$$= \frac{[\text{U-235}]_{\text{now}}\{\exp(\lambda_{235}t_a) - 1\}}{[\text{U-238}]_{\text{now}}\{\exp(\lambda_{238}t_a) - 1\}}. \qquad (8.7)$$

The ratio $[\text{U-235}]_{\text{now}}/[\text{U-238}]_{\text{now}}$ is known to be 7.2527×10^{-3} so equation (8.7) is a relationship involving only measured concentrations of lead isotopes. Writing

$$m = 7.2527 \times 10^{-3}\frac{\exp(\lambda_{235}t_a\} - 1}{\exp(\lambda_{238}t_a) - 1} \qquad (8.8)$$

and dropping the subscript 'now', equation (8.7) can be rearranged to give

$$\frac{[\text{Pb-207}]}{[\text{Pb-204}]} = m\frac{[\text{Pb-206}]}{[\text{Pb-204}]} + \frac{[\text{Pb-207}]_0}{[\text{Pb-204}]_0} - m\frac{[\text{Pb-206}]_0}{[\text{Pb-204}]_0}. \qquad (8.9)$$

Plotting $[\text{Pb-207}]/[\text{Pb-204}]$ against $[\text{Pb-206}]/[\text{Pb-204}]$ gives a straight line of slope m which is a function of t_a, which can then be

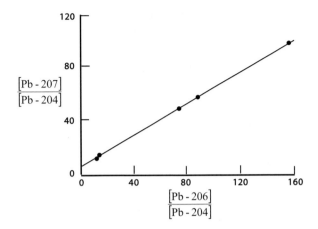

Figure 8.6   A lead–lead isochron for a meteorite specimen.

determined. The intercept on the y-axis is a function of m, which is determined, and two unknown initial lead concentration ratios that cannot be individually determined. A lead–lead isochron of a meteorite with age 4.543 billion years is shown in Figure 8.6.

8.7 The Age of the Earth

Ideas about the age of the Earth have varied greatly over the time that humankind has considered the question. The Greek philosopher Aristotle thought that the world had existed eternally, so that its age was effectively infinite. By contrast the Roman poet Lucretius believed that the world was of recent origin since there were no records going back beyond the Trojan wars.

However, the first quantitative assertion concerning the age of the Earth is found in the Jewish calendar that begins in 3760 BCE and is taken to denote the time of creation when everything in the Universe, animate and inanimate, was created. This creation date was refined in 1650 by the Irish Archbishop of Armagh, James Ussher (1581–1656), who followed the chronology of the bible and calculated that the world was created on October 23rd 4004 BCE.

The English astronomer and friend of Isaac Newton Edmond Halley (1656–1742) made a contribution of a more scientific nature by considering the salinity of the oceans. The water that forms them, flowing in from rivers and originally falling as rain, is initially fresh water, and over the course of time dissolved salt from land-based sources would enter the oceans. It could be deduced from the degree of salinity that the Earth had to be of considerable age but it could not be of extremely great age, otherwise the oceans would be saturated with salt, which they are not. Although Halley's approach was purely qualitative, in 1899 an Irish geologist, John Joly (1857–1933), used the idea to estimate the age of the Earth and came up with a value in the range 80–90 million years.

Following Halley, the next contribution of a scientific nature came from the Scottish geologist James Hutton (1726–1797), often referred to as the 'Father of modern geology'. He was the first to put forward the concept of *uniformitarianism*, the principle that geological changes come about by natural processes — such as erosion and sedimentation — operating over long periods of time. This was in contrast to a previous idea of *catastrophism* which believed that the world was shaped by catastrophic events, such as the biblical flood. A natural consequence of uniformitarianism is that long periods of time were required to shape the world as we see it today. However, because the world has been changing its form constantly, Hutton concluded that it was impossible to determine its age, although it had to be of much greater age than the biblical account would suggest.

James Hutton

The next noteworthy contributor to the problem of the Earth's age was the British geologist Charles Lyell (1797–1875), who was a strong supporter of Hutton's ideas. He was the author of *The Principles of Geology*, which was the definitive work on geology for the whole of the Victorian period. In it, he introduced the idea of *stratigraphy*, the study of strata in rocks and the fossils within them to determine the relative ages of rock formations. Although his approach could not give a precise estimate of the age of the Earth, he did propose on the basis of his work that the age of the Earth was more than 300 million years.

Charles Lyell

In 1862, the British physicist William Thomson (later Lord Kelvin, 1824–1907) refined a method first suggested by the French scientist Georges-Louis Leclerc, Comte de Buffon (1701–1778). This was based on the idea that when the Earth formed it was completely molten and that its age could be determined by estimating the time it would take to cool to its present state. Buffon's estimate was 75,000 years but, by a more refined calculation, Thomson obtained an estimate in the broad range of 24–400 million years. What neither Buffon nor Thomson realized was that there is an internal source of heat within the Earth — radioactivity. It was this phenomenon that would provide a definitive answer to the question 'How old is the Earth?'

Lord Kelvin

The first age estimates using radioactivity were made by the New Zealand physicist Ernest Rutherford (1871–1937) and the American chemist Bertram Boltwood (1870–1927): When uranium decays it releases α-particles (helium nuclei), which could be regarded as a daughter-product that is trapped in the interstices of the rock. However, the rock they analysed was permeable to helium which gradually escaped from the rock, so their age estimates were quite poor underestimates. Eventually they used the uranium-to-lead method and obtained much better results, which made it clear that all previous work gave gross underestimates of the Earth's age.

Ernest Rutherford   Bertram Boltwood

The method has been greatly improved over the years and the age of the oldest rocks on Earth using zircons coming from Jack Hills in Western Australia is about 4.4 billion years. Most rocks on Earth are much younger. The Earth is tectonically active and the surface in most regions is constantly turning over, with surface material moving into the interior and interior material moving to the surface. The high accuracy of present age estimates is due to an instrument called a *mass spectrometer*, which will now be described.

8.8 The Mass Spectrometer

The working of a mass spectrometer depends on the way that charged particles behave in electric and magnetic fields. Figure 8.7 shows the motion of a negatively charged particle when it moves through a region with an electric field. The force on the particle is proportional to the product of the charge on the particle q and the strength of the electric field E, i.e.

$$F_E = qE. \tag{8.10}$$

The motion of a charged particle in a magnetic field is illustrated in Figure 8.8. The magnetic field, which is pointing into the page (the circle with contained cross representing the rear view of an arrow) and is of strength B, is perpendicular to the motion of the particle moving at speed v. The force F_B on the particle with charge q is

$$F_B = qBv \tag{8.11}$$

and is perpendicular to both B and v (Figure 8.8a). This force curves the path of the particle and if the magnetic field is uniform over the region of

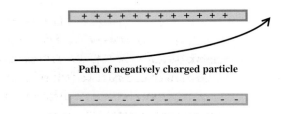

Figure 8.7 The path of a negatively charged particle in an electric field.

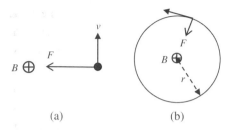

(a)   (b)

Figure 8.8   (a) The force on a moving charged particle in a magnetic field. (b) Motion in a circle.

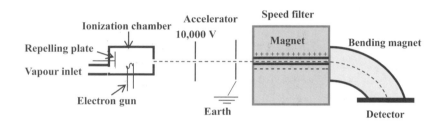

Figure 8.9   A schematic of a mass spectrometer.

its motion, it will move on a circular path, as shown in Figure 8.8b. The radius of this path is found by equating the centrifugal force outwards due to its circular motion with the centripetal force inwards due to the magnetic field, giving

$$\frac{mv^2}{r} = qBv$$

or

$$r = \frac{mv}{qB}. \tag{8.12}$$

We now show in Figure 8.9 the features of a mass spectrometer.

The sample material is vaporized and fed into the ionization chamber. The individual atoms in the chamber are ionized by electron bombardment, the resulting ions having a positive charge of magnitude equal to that of the ejected electron. This is then repelled by a positively charged repelling plate and emerges from the ionization chamber in the form of a fine stream. They then pass through the accelerator which gives a more uniform speed distribution. To further improve the uniformity in the speed of the ions they pass through a narrow aperture in a speed filter. In the aperture there is an electric field and a magnetic field, producing forces in opposite directions. The net force on a particle is zero when F_E equals F_B or when $v = E/B$. Only the particles with that speed will pass through the filter on a straight path, others being deviated up or down.

The next stage is when the particles pass through the bending magnet, and the degree of bending will depend on the mass of the particle, lighter particles being more deflected than heavier ones. Detectors can be of various types, but they all, directly or indirectly, measure the total charge carried by the ions they receive. There can be a single detector with a small aperture and then, by varying the field produced by the bending magnets, beams of isotopes of different mass can be directed into the detector.

This description of a mass spectrometer is very simplified but gives the essence of its mechanism.

8.9 The Age of the Solar System

There are alternative models for the formation of the Solar System, differing in their interpretation of the presence of asteroids. In one interpretation asteroids are the initial building blocks from which planets formed with the unused asteroids being those present today. In the other interpretation the planets were produced from the collapse of dusty gas clouds and asteroids are the product of the break-up of planets due to a collision. However, for present purposes the exact model for their formation is of no consequence since both have them formed within a few million years of the formation of the system and the ages of the meteorites, fragments from asteroids, will give an indication of the age of the Solar System within the accuracy limits of the methods used.

The age determinations from meteorites vary slightly but the consensus age from all the observations is 4.57 billion years, similar to that indicated by the lead–lead isochron shown in Figure 8.6. This is somewhat greater than the age of the oldest rocks on Earth, those from Jack Hills, but the significance of this is not clear, although it probably indicates that the Jack Hills zircon was produced some time after the Earth formed.

While most meteorites are found to be about 4.5 billion years old, close to the age of the Solar System, a few stony meteorites are much younger, with ages around one billion years. They contain gas trapped in cavities, and when this gas is analysed it is found to be similar to the composition of the atmosphere of Mars. These meteorites are of three types, *shergottites, nakhlites and chassignites,* named after the places where the prototypes were found, and together they are referred to as *SNC meteorites.* It is believed that these meteorites originate from the planet Mars, presumably fragments ejected from the surface following an asteroid collision. They indicate the dates when the material became solid, which is consistent with the idea that there was volcanism on Mars relatively recently, i.e. about one billion years ago.

Ages on Earth

Chapter 9

The Age of the Continents

9.1 The Mobile Earth

The understanding of the Earth in the distant past was that its general form is static and unchanging with time. There may be occasional earthquakes, volcanic eruptions, floods or other catastrophes that make minor local changes but, by and large, the Earth at any time is as it has always been since it first formed. This view was first challenged in the sixteenth century. In 1564, a Dutch cartographer, Abraham Ortels (1527–1598; Figure 9.1a), produced his first reasonably accurate map of the world as it was then known (Figure 9.1b). He noticed that the east coast of South America and the west coast of Africa have similar shapes and could fit together like two pieces of a jigsaw puzzle. In 1596, he suggested that at one time these two continents had been joined together but had been pushed apart by 'earthquakes and floods'.

This fit of the continents was regarded as just an interesting coincidence and attracted little scientific interest for the next 316 years, but in 1912 a German geologist, Alfred Wegener (1880–1930), proposed in a published paper that continents moved, an idea he expressed as 'die Verschiebung der Kontinente' which translates into English as *continental drift*. Wegener suggested that the continents move through the ocean crust in the way that a ploughshare moves through the Earth and this idea was derided by the majority of geologists. They could not envisage any forces that could conceivably move whole continents around in this way. Later, the evidence for continental drift became irrefutable but even so many geologists, in particular the leading British geologist, Harold Jeffreys (1891–1989), never accepted the idea.

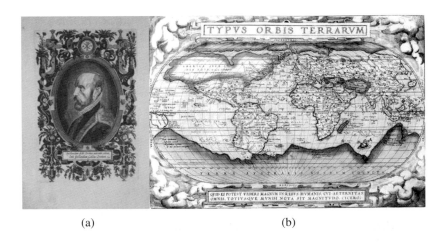

(a)   (b)

Figure 9.1   (a) Abraham Ortels. (b) Ortels's world map, 1570.

Alfred Wegener

The Wegener model for the way that continents move was quite untenable but nevertheless, continents do move and we shall now consider the evidence for this.

9.2 The Evidence for Continental Drift

Although Wegener's ideas of continental drift seemed outlandish, there arose a body of evidence that indicated that what he proposed is true. Fossils in different continents showed the same plants and animals of the same age in widely different locations, something that would be impossible if there were no link between these locations. As the fossil evidence increased, it became possible to determine how the continents could have fitted together to explain the observations. A proposed arrangement of the southern continents 200 million years ago is shown in Figure 9.2. The coloured bands show the locations of fossils of the same age and type and it can be seen how they link the continents together into a single land mass, given the name *Gondwana*. An equally important indication is the similarity of rock types in the present continents corresponding to contiguous regions of Gondwana.

There were many other indications that continents must have moved through large distances. For example, there is the presence of coal and fossils of tropical plants in Antarctica. Another indication is striations in the surfaces of the southern continents, due to the motions of giant glaciers, which line up when the continents are assembled into Gondwana. With the continents as presently placed the motion of these glaciers tend to run from the equator southwards, indicating clearly that the continents were once placed differently with respect to the Earth's spin axis.

Putting all the scientific evidence together gives a picture that 200 million years ago the land masses were arranged as in Figure 9.3. The present northern continents are linked as the land mass *Laurasia* and the combination of Gondwana and Laurasia is called *Pangaea*, a word derived from Greek meaning 'all the Earth'. There is no agreement about the precise arrangement of land masses in the distant past. Some interpretations of Pangaea give a lesser separation of the Americas by moving North America further south and this reduces the distinction between the two continents Laurasia and Gondwana.

9.3 The Mechanism of Continental Drift

By the 1950s, the idea of continental drift was generally accepted but what was missing was a credible theory for the way it occurred — Wegener's ploughshare model was clearly not possible. To understand the actual mechanism we need to consider the structure of the outer regions of the Earth, from the surface to a depth of several hundred kilometres.

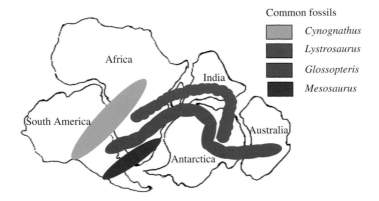

Figure 9.2   Regions of fossils of similar species in Gondwana. *Cynognathus* and *Lystrosaurus* are land reptiles from about 250 million years ago. *Glossopteris* is a type of fern and *Mesosaurus* a fresh-water reptile.

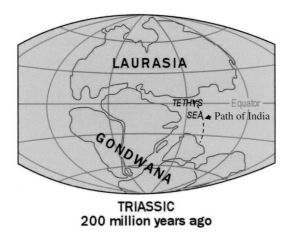

Figure 9.3   Pangaea, the land mass of the Earth 200 million years ago (US Geological Survey — USGS).

The solid surface layer of the Earth known as the *lithosphere* (from the Greek *lithos*, meaning stone) consists of the low-density rocks that constitute the crust of the Earth plus a lower region of denser solid rocks forming part of the *mantle* of the Earth, the silicate region surrounding the central metallic core. Below the lithosphere, as the temperature increases the mantle eventually becomes fluid, but between

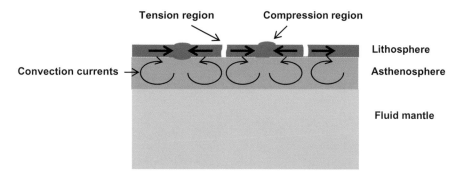

Figure 9.4   The near-surface structure of the Earth, convection currents and forces on lithosphere.

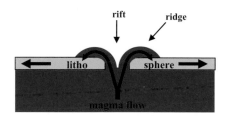

Figure 9.5   Creation of new lithosphere in the gap when the exiting lithosphere is torn apart.

the completely fluid region and the lithosphere the mantle, although virtually solid, can flow like a very viscous liquid; this region is known as the *asthenosphere*. Slow convection currents in the asthenosphere bring up heat from below to enhance the rate of Earth cooling and they also apply drag forces on the lithosphere, tending to tear it apart in some regions and crush it together in others. Figure 9.4 shows the near-surface structure of the Earth, the convection currents and the forces on the lithosphere.

In regions where the lithosphere is being pulled apart magma wells up from the Earth's mantle to fill the gap. Eventually, this solidifies to form a new section of lithosphere which continues to move outwards. A ridge forms on either side of the rift (Figure 9.5).

Running the full length of the Atlantic Ocean there is a crack in the Earth's surface, called the *mid-Atlantic Ridge*, which is formed in

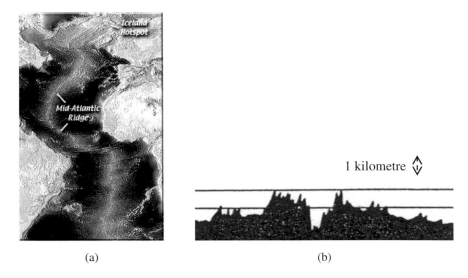

(a)   (b)

Figure 9.6   (a) The mid-Atlantic Ridge (USGS). (b) A cross-section of the ridge
(vertical and horizontal scales are different).

this way (Figure 9.6a). A typical cross-section of the ridge is shown in
Figure 9.6b.

The new lithosphere forming in the region of the mid-Atlantic
Ridge causes North America and Europe to move further apart; satellite
measurements show that the Atlantic Ocean is widening by 2.5 cm
each year, which may not seem much but amounts to 2,500 km in 100
million years, just 2.2% of the age of the Earth. This illustrates how
over geological time periods the continents can move. However, the
Earth is not expanding overall, so since new surface is being created in
the region of cracks, it must be being destroyed elsewhere.

The loss of surface area occurs in the compression regions of
the Earth. There are two ways this can happen. The first, shown in
Figure 9.7, is when the compression occurs within the continental
crust and the material being pushed together is of the same density
and type. The lithosphere is then crushed and buckles both upwards
and downwards to give mountain formation. In Figure 9.3, the land
mass destined to become the Indian sub-continent is shown moving
upwards towards the southern flank of Laurasia. It is the force of the

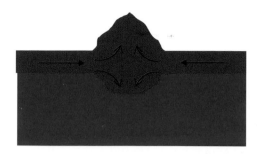

Figure 9.7 Compressive forces on the lithosphere leading to mountain formation.

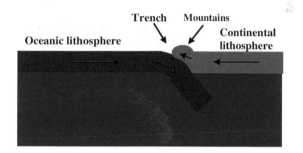

Figure 9.8 Subduction of an oceanic plate below a continental plate giving trench and mountain formation.

collision between these landmasses that created the Himalayas, which are still rising by one centimetre per year as India slowly continues its northward motion at a rate of a few metres per century. This collision also caused the formation of the Tibetan Plateau with an average elevation of 4.5 km. Crushing and buckling of the surface leads to a loss of surface area, but a comparatively modest loss.

The second outcome of compressive forces is much more effective in reducing the surface area of the Earth. It occurs when the compression is at the boundary between continental crust, with a density about 2,700 kg m^{-3} (kilograms per cubic metre), and the oceanic crust, with a higher density of 2,900 kg m^{-3}. What happens then is shown in Figure 9.8; the less dense continental crust rides over the denser oceanic crust which is thrust downwards towards the mantle — a process known as *subduction*. This is a very effective mechanism for reducing surface area. It produces deep off-shore trenches and the formation of mountains on the continental side. The subduction of Pacific ocean-crust material under the western coasts of North and South America is the agency that forms the Andes and Rocky Mountains running along those coasts and is also responsible for these regions being prone to earthquakes.

Sometimes, the compression is within an ocean region and again subduction is the outcome. One side or the other moves downwards and, since the density of oceanic crust is similar to that of the mantle, there is little buoyancy resistance to this subduction taking place. When this happens, a trench is formed as shown in Figure 9.8; an example of such a feature is the Marianas Trench in the Pacific Ocean, with a depth at its southern end in the Challenger Deep of almost 11,000 metres. This is greater than the height of Mount Everest above sea level — 8,854 metres.

9.4 Plate Tectonics

We have been describing regions of the Earth's crust where along lines of collision some Earth-modifying processes are taking place. This leads to the surface of the Earth being divided into *tectonic plates*, regions that retain their overall identity but are constantly modified by interactions with neighbouring plates in one or other of the ways that have been described. Figure 9.9 shows the plate structure that covers the Earth's surface. In it the plate edges can be seen that define the mid-Atlantic Ridge (black circles) and also the Pacific Plate, the eastern boundary of which is being thrust under North and South America. The red circles indicate the 'ring of fire' which runs round the edge of the Pacific Ocean covering New Zealand, Indonesia, the Philippines, Japan, Korea and the west coast of the Americas from Alaska to the southern part of Chile. This region accounts for more than 80% of the major earthquakes that occur.

The combination of mountain building, formation of new surface at tensional features and subduction maintains a constant surface area on the Earth but also results in the motions of continental plates relative to each other.

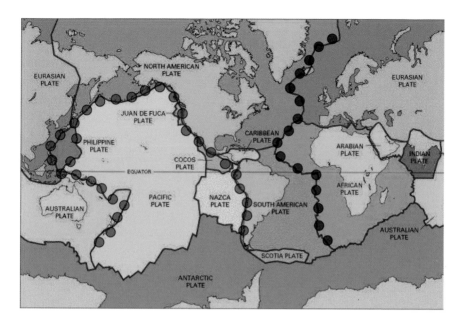

Figure 9.9   The tectonic-plate structure of the Earth's lithosphere. Black circles mark the location of the Mid-Atlantic Ridge and red circles that of the 'ring of fire', regions of active volcanism (USGS).

9.5 Defining the Age of Continents

This chapter has the title 'The Age of the Continents' and the question arises of what is meant by 'age' in this context. For animate entities age is defined from the time of birth, but for the ages of inanimate entities it is necessary to define a time of 'birth'. The age of a rock was defined in Chapter 8 as the time from when it became a closed system, neither losing material to nor gaining material from outside. In one sense, the continents have existed from the time that a lithosphere formed and broke up into tectonic plates, but the distinct entities that we now recognize as continents would not have been evident then. The surface of the Earth is mobile and changes greatly on a geological timescale. Figure 9.10 shows how it has changed in the last 250 million years; 100 million years ago the arrangement of surface features was very different from that now and will again be different 100 million years in the future.

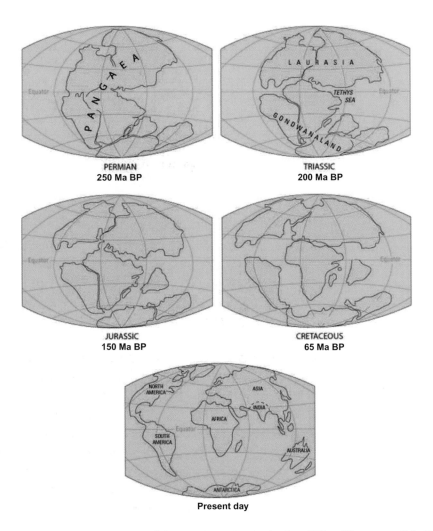

Figure 9.10   Movements of the continents over the last 250 million years (Ma BP means 'million years before present time'; USGS).

If we define the 'birth' of the continents as that time when their distinctive and separated shapes became apparent regardless of their location on the globe, then a perusal of Figure 9.10 suggests that their age is somewhere in the broad range of 65–150 million years. It is interesting to note that when the first reptiles roamed the Earth during the Triassic period, to be followed by dinosaurs and the first mammals, it was a very different Earth which they occupied.

Chapter 10

The Ages of Fossils

10.1 What is a Fossil?

A fossil is a recognizable residue of some living organism that lived in the distant past. The organism can be as small as a bacterium or as large as a dinosaur weighing several tonnes or some of the largest trees that have ever existed on Earth. The form of the fossil depends on the nature of the organism that produced it and the environment in which it died.

For invertebrate animal and vegetable life consisting of soft tissue that would completely decay, the fossil consists of some impression of the object in the material in which it is embedded. Thus, if a dead soft-tissue organism is buried in a sediment that dries out and sets hard before the decay process is complete, then, when the sedimentary material is compressed into a rock, it will contain a cavity showing the shape of the organism. Such a fossil is called a *mould*. Sometimes, the cavity will fill with some different material than the surrounding rock and the resulting fossil is termed a *cast*. Life began on Earth about 3,500 million years ago. The first form of life was single-celled organisms similar in appearance to bacteria. There seemed to be some bacteria, as found in hot springs in Yellowstone Park, which had apparently adapted to live in very hostile environments — at high temperature, in conditions of high alkalinity or high acidity or even within oil deposits. Then, in 1977, the DNA of these organisms was studied and found to be quite different from that of bacteria, so clearly they were not related. They were named *Archaea* and they were possibly the first life forms in the hostile environment of the early Earth, although they would have

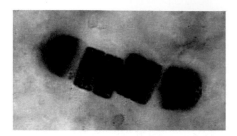

Figure 10.1   A fossil of blue-green algae, 3.5 billion years old.

Figure 10.2   Stromatolite from Glacier National Park, Montana.

been quickly followed by bacteria as conditions improved. An early form of bacterium was *cyanobacteria*, also known as *blue-green algae*, a fossil of which is shown in Figure 10.1. It is an important organism because it carries out the process of *photosynthesis*, which transforms carbon dioxide and water into oxygen and cellulose, and the early establishment of an oxygen component in the Earth's atmosphere owes much to these bacteria.

Cyanobacteria formed large one-dimensional mats, usually in shallow water. These became embedded in sediments, which eventually formed sedimentary rock which contained the fossil imprint of the bacterial mat in the form of a mould, known as a *stromatolite* (Figure 10.2).

The earliest life on Earth developed in the sea and the first plants and animals became established on land about 420 million years before the present time (Ma BP). With buoyancy to support their weight, many of the invertebrate creatures in the sea attained considerable size.

Figure 10.3 A large ammonite fossil.

Figure 10.4 A petrified log from the Petrified Forest National Park, Arizona.

One species, which developed around 400 Ma BP, were *ammonites*, molluscs that in the later periods of their existence grew quite large. Figure 10.3 shows a fossil cast of a particularly large specimen.

For vertebrates and ligneous material, fossils are formed by partial or complete replacement of the original material — bone or wood — by minerals, a process known as *petrifaction*. An example of a petrified log coming from the Petrified Forest National Park in Arizona is shown in Figure 10.4. None of the original wood is present but the form of the original tree is faithfully reproduced in stone.

Fossils of vertebrates are sometimes in the form of moulds; Figure 10.5 shows a fossil mould of *Archaeopteryx*, a flying dinosaur

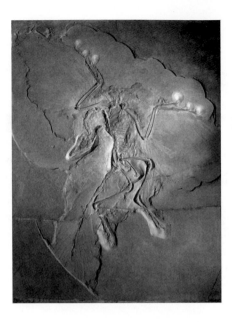

Figure 10.5   A fossil of archaeopteryx, a flying dinosaur (Museum für Naturkunde, Berlin: H. Raab).

from about 150 Ma BP, no bigger than a pigeon and representing the transition between dinosaurs and true birds.

Larger dinosaurs are often excavated in the form of complete or almost complete skeletons and many magnificent specimens are to be found in natural history museums all over the world. Figure 10.6 shows a fossil skeleton of the carnivorous dinosaur *Allosaurus*, some 11 metres in length, from the period between 200 and 145 Ma BP.

We have now surveyed a range of fossil types, shown examples of fossils and assigned approximate ages to them, but what we have not done is to explain how those ages are determined, something we must now address.

10.2 Stratigraphy

When the side of a cliff or canyon is observed, it is often seen that it shows a layered structure. Figure 10.7a shows a cliff face at Lyme Regis in Dorset on the south coast of England, part of the Jurassic Coast, the

Figure 10.6   The fossilized remains of an *Allosaurus* (San Diego Natural History Museum).

(a)   (b)

Figure 10.7   (a) Strata in the cliffs at Lyme Regis, Dorset, UK (Michael Haggs). (b) Strata in the sides of the Grand Canyon.

location of a large number of fossil finds. An even more spectacular layered structure is evident in the sides of the Grand Canyon in the USA (Figure 10.7b). These layers are called *strata* and the study of their structure is called *stratigraphy*.

These layers were laid down horizontally in sequence so that newer layers overlay older ones. However, the horizontal layering is sometimes disturbed by upheavals that can tilt the strata, either to a small extent as shown in Figure 10.8a, part of the Dakota Hogback, a long ridge

(a)   (b)

Figure 10.8   (a) Tilted strata at Dakota Hogback. (b) Highly tilted strata in Atigun
Valley.

at the eastern fringe of the Rocky Mountains, or to a large extent,
sometimes to be almost vertical, as seen in the Atigun Valley in Alaska
(Figure 10.8b).

10.3 Fossil Ages Determined from Surrounding Rocks

Some fossils will contain the radioisotope carbon-14 and can be dated
directly (Chapter 12), but this can only be used for ages up to 70,000
years, which in geological terms is the recent past. Here we shall be
concerned with much greater ages.

When fossils are found in different strata in a particular environment, then comparative dating of the fossils is possible, at least to give
the order in which the remains were deposited. If the time of deposition
of the rock in which a fossil is found is known, then so is the age of
the fossil. Conversely, if the age of the fossil is known, then so is the
time of deposition of the stratum in which it is found. You will notice
that it is the 'time of deposition' of the rock that is referred to and
not its age. Many layers of sedimentary rock have formed from eroded
fragments of igneous and metamorphic rocks, so any age determined
will be from when the fragment sources became closed systems and
not when the sedimentary rock layer was established — which is what
is needed to date fossils. However, if the fossil is embedded in either an
igneous or metamorphic rock, then the age of the rock will also date
the fossil. Another situation in which absolute dating is possible is when

the sediment consists of volcanic ash derived from molten material, so the time measured will give the age of the deposit.

Another aid to dating rocks is *palaeomagnetism* — using the phenomenon of the occasional reversals of the Earth's magnetic field, when the north and south magnetic poles switch directions. This is recorded in many rocks which contain material that can be magnetized. A molten rock cannot be magnetized nor can a solid rock above a certain temperature known as the *Curie temperature* — which is 858 K for magnetite, Fe_3O_4, a rock that can be strongly magnetized. Once the rock has cooled below the Curie temperature, then it will retain that direction of magnetization no matter what the subsequent changes in the magnetic environment. One way of detecting magnetic field reversals is by measuring the direction of magnetization of the sea floor on both sides of the mid-Atlantic Ridge (Section 9.3). When the emitted magma cools below the Curie temperature it records the direction of the prevailing terrestrial magnetic field. It is then pushed outwards by new magma and when the magnetic field reverses, which happens very quickly, there is a sudden change in the direction of magnetization of the seabed rock. Figure 10.9 gives a representation of magnetic recordings on either side of the mid-Atlantic Ridge.

ridge

Figure 10.9   A schematic picture of magnetization of the seafloor on either side of the mid-Atlantic Ridge. Black stripes indicate magnetization in the present direction of the Earth's magnetic field. White stripes indicate magnetization in the opposite direction.

The reversals take place randomly, with intervals varying between a fraction of a million years and tens of millions of years. Through studies of magnetic-field directions in large numbers of rocks that can be radioactively dated, the times of reversal events have been found. While it is impossible to assign an orientation of the magnetic field to a particular date, if the orientation of one dated stratum can be linked to a particular orientation then the ages of a whole sequence of neighbouring strata can be determined from the orientation of the magnetization within them.

From the dating of fossils of various kinds in favourable situations — such as when they are buried in volcanic ash or by using palaeomagnetic information — a body of knowledge has been built up concerning the arrival and extinction of different species. This kind of information can be used to refine the dating of fossils using what is known as the *principle of faunal succession*. This is illustrated in a simplified form in Figure 10.10. It illustrates the existence spans of the fossils of five organisms, A, B, C, D and E, which are in strata that it has not been possible to date. The presence of A alone would allow a possible age range of between 49 and 108 Ma BP, but the neighbouring presence of the fossils of C, D and E indicate that the age of the stratum, and hence the fossils within it, must be in Timespan 1, i.e. between 96 and 108 Ma BP.

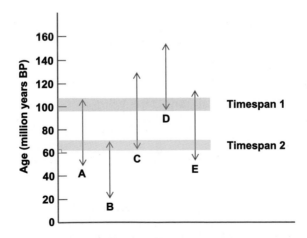

Figure 10.10   Using the principle of faunal succession to refine the dates of fossils.

Similarly, if fossils A, B, C and E existed in a stratum, then their ages would be restricted to Timespan 2, i.e. the range 63–71 Ma BP. When there is a large number of fossil organisms present then this technique can determine the ages of strata with a small range of uncertainty.

There is an interesting geological feature that is observed all over the world. At 65½ Ma BP, the Earth was struck by an asteroid in the Yucatan Peninsula in Mexico. It was about 10 km in diameter and it released an amount of energy equivalent to that of 100,000 large hydrogen bombs. This caused devastation over the whole globe, with possible effects including the production of worldwide fires and dust blanking out radiation from the Sun over an extended period. This led to the extinction of dinosaurs — the herbivores because of the lack of vegetable food and the carnivores because their prey died out. The collision, known as the KT event (Section 11.5.3), is recorded on Earth by a thin worldwide stratum rich in the element iridium. This element is not very abundant on Earth but is more abundant in meteorites and presumably their parent bodies, asteroids.

10.4 The Use of Electron Spin Resonance (ESR)

The individual electrons in an atom have the property of being tiny magnets, with a north pole and south pole. In most atoms, the existence of this magnetic property of the electrons is not evident because pairs of electrons pair up pointing in opposite directions so their magnetic effects cancel each other out. However, there are some atoms in which there is an unpaired electron and in such cases it is possible to detect the presence of the electron magnet. Another circumstance in which there are unpaired electron magnets is when a *free radical* is formed. An example of free-radical formation is obtained by exposing a water molecule to high energy radiation that causes the molecule to break up into two parts — a free-radical hydrogen atom, H, and a hydroxyl free radical, OH, viz.

$$H_2O \to H + OH. \tag{10.1}$$

Atoms form molecules by either donating electrons to or accepting electrons from each other or by sharing electrons; when a free radical is formed it has one or more electrons that it would like to share or donate, or it is possibly a candidate for receiving a donated electron, so it is chemically very reactive and will combine with something else given the opportunity to do so. One environment where free radicals exist is in the tails of comets. The Sun's radiation breaks up molecules to form free radicals but the density in the comet's tail is so low that it takes a long time before the free radical comes into contact with something with which to combine. When a fossil is buried it is exposed to natural radiation in its environment and this produces free radicals within material such as bone and teeth, which are trapped within the material and are unable to combine chemically with other materials. The concentration of free radicals and hence of unpaired electrons increases with time so that measuring that concentration gives a measure of age.

When a magnet such as a compass needle is placed in the Earth's magnetic field, its north pole points towards the north because the Earth, considered as a magnet, has its *south pole* in the northern hemisphere. However, the electron magnets behave rather differently in a magnetic field. Some line up as would be expected of a compass needle (parallel state) while others line up in the completely opposite direction (antiparallel state); there is a slight preponderance in the parallel state. The energy associated with these two states is different, with the lower energy, E, for those in the parallel state being given by

$$E = -CB, \tag{10.2}$$

where C is a known constant for an electron and B is the strength of the magnetic field. For the opposite direction the energy is CB so the difference in energy between the two states is $2CB$. The idea of a photon was explained in Section 4.3.1; electromagnetic radiation of frequency v exists in packets of energy hv. If the electron magnets in a magnetic field of strength B are exposed to electromagnetic radiation of frequency v such that

$$hv = 2CB, \tag{10.3}$$

then two things can happen. Firstly, the photon energy can be absorbed to flip electrons from the parallel to the antiparallel state or, alternatively, they can stimulate those in the antiparallel state to flip to the parallel state with the emission of another photon of frequency v, giving two photons in place of the original one. For any particular electron the likelihood of the two kinds of flip are equal but, since there are more electrons in the lower-energy state that absorb photons than there are in the higher-energy state creating new photons, the net effect is that some of the radiation is absorbed.

In a typical ESR experiment the specimen is placed in the field of an electromagnet such that the field can be changed by varying the current. The specimen is exposed to microwave radiation of frequency v and the reflected radiation is fed into a detector. The magnetic field is scanned so as to pass through the strength of field that would satisfy equation (10.3) and as it passes through that value the intensity of the reflected radiation falls, showing that the radiation is being absorbed by the specimen (Figure 10.11).

The magnitude of the absorption will depend on the concentration of free electrons in the specimen and, in its turn, that concentration will depend on the total exposure of the specimen to radiation, which depends on the exposure time. This method can only be used for fossil ages less than about 10 million years, since there is a saturation of sites in which the free radicals can occur and beyond a certain radiation dose there is little increase in the concentration of free electrons.

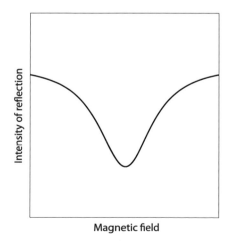

Figure 10.11   Output from an ESR experiment.

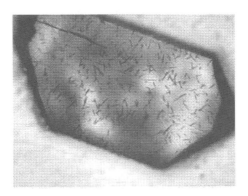

Figure 10.12   Fission tracks in a zircon sample.

10.5 Fission Track Dating

This uses the fact that the isotope uranium-238 undergoes spontaneous fission where the atom breaks into two comparable parts, a process that is distinct from radioactivity and has a very long half-life, 8.2×10^{15} years. The heavy particles produced by the fission move through the mineral causing damage by multiple collisions until it comes to rest. When the surface of the specimen is polished and etched, the damaged regions can be seen because they are weaker and more strongly etched than the undamaged surroundings (Figure 10.12). Assuming that the uranium is distributed uniformly throughout the sample and knowing the rate of fission, a measurement of the number of fissions within a region allows the time the region has remained undisturbed (i.e. its age) to be estimated.

The method is especially valuable because the fission tracks fade by an annealing process at higher temperatures and the distribution of the number and the quality of tracks in different mineral samples can allow their temperature histories to be inferred.

10.6 What is Learned from the Dating of Fossils?

The field of *palaeontology*, the dating of fossils by the methods that have already been described and other methods that will be described in Chapter 12, has given a body of knowledge that has enabled an evolutionary and dated sequence to be built up of the forms of life that have existed, and still exist, on Earth from bacteria to mankind. It is a fascinating story, linked with Darwin's evolutionary theory of 'The Survival of the Fittest', so, as environmental conditions have changed, organisms have either adapted or become extinct to be replaced by other species better able to survive. In the following chapter we give an outline of this story.

Chapter 11

The History of Life on Earth

The time from the formation of the Earth to the present can be divided into *eras*, *periods* and *epochs*. An era is a time within which changes of a dominant life form occur — from dinosaurs to mammals for example. Eras are subdivided into periods in which changes of life form are considered in a more detailed way. Finally, as we approach the present time, periods are subdivided into epochs in which events take place, such as the developments of the first primates to man, which relate to life as it is today. Most changes are gradual through the slow process of evolution, but occasionally there are abrupt changes due to catastrophes such as the KT event (Section 10.3) that caused mass extinctions and also due to mutations, chance changes of the DNA structure of an organism that produce some radically different form of life over a short timescale.

11.1 The Hadean Era (4,500–3,800 Ma BP)

Within this era, the Earth's temperature fell, allowing a solid crust and liquid water to form. Strong convection in the asthenosphere created tectonic plates and continental drift occurred at a high rate. A thick atmosphere, mostly of carbon dioxide but with some nitrogen, was present. A small amount of oxygen was probably being produced by the break-up of water vapour through solar radiation. The OH radicals, produced as shown in equation (10.1), combined together as

$$OH + OH \rightarrow H_2O + O. \tag{11.1}$$

Two oxygen atoms would then combine to give an oxygen molecule, O_2, or sometimes three would come together to produce an ozone molecule, O_3. The ozone layer in the present Earth is important because it absorbs ultraviolet radiation from the Sun which is harmful to living organisms. It is possible that a tenuous ozone layer was forming in this era.

11.2 The Archaean Era (3,800–2,500 Ma BP)

Although in this era the Sun was less luminous than now, the combination of residual heat in the Earth and more radioactive heating — at about three times the present rate — would have held the temperature to about its present level. Additionally, due to the carbon dioxide atmosphere there would also have been a large greenhouse effect. The first life appeared, probably Archaea, but cyanobacteria would also be present, thus beginning the generation of oxygen from that source.

11.3 The Proterozoic Era (2,500–543 Ma BP)

The first oxygen produced by cyanobacteria was mostly absorbed by oxidizing surface materials, but during this era oxygen began to accumulate in the atmosphere. This oxygen enabled the first *eukaryote* to form — organisms with genetic material contained in a nucleus bounded by a membrane. Multi-celled organisms occurred in the form of algae and soft-bodied worm-like creatures. The cyanobacteria mats that produced stromatolites also formed in this era (Figure 10.2).

11.3.1 *The Ediacaran period (600–543 Ma BP)*

The fossils of soft-bodied creatures first appear in this period (Figure 11.1). At the beginning of the period, there is a layer of carbonates deficient in the isotope carbon-13 on top of glacial deposits. It has been postulated that these glacial deposits were produced when the whole Earth was frozen, giving rise to what has been called 'snowball Earth' an event followed by the rapid evolution of life. The significance of the carbon-13 deficiency is that organisms living in the oceans that use the

Figure 11.1   A fossil from the Ediacaran period (British Geological Survey).

process of photosynthesis, such as algae, preferentially use the isotope carbon-12, so that organic remains in the oceans would be deficient in carbon-13. Thus, the organic component of sediments in the oceans would show a deficit of carbon-13 and this would be left behind as a deposit when the frozen water melted.

11.4 The Paleozoic Era (543–251 Ma BP)

Life developed rapidly in this era, producing most types of life that have ever existed, with the exception of mammals and flowering plants. Occasional events occurred that led to the loss or decimation of some species and the era ended with a major extinction of species that led to new types of creatures dominating the Earth.

11.4.1 *The Cambrian period (543–488 Ma BP)*

In the Cambrian period, life became more varied, more abundant and more complex — the so-called *Cambrian Explosion*. The reason for this may have been an increase in atmospheric oxygen, allowing a higher metabolic rate that could support larger and more complex organisms. For the first time predatory life appeared; previously food had been obtained from decaying organic material or by developing a symbiotic relationship with photosynthesizing algae.

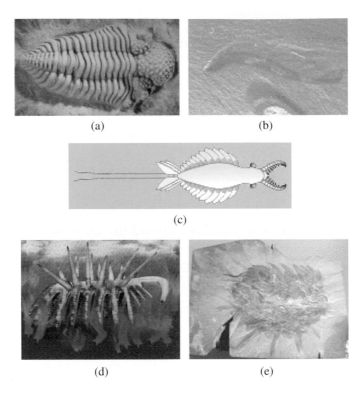

Figure 11.2   A selection of creatures from the Cambrian period. (a) A trilobites fossil. (b) A *Pikaia* fossil. (c) A drawing of an *Anomalocaris*. (d) An artist's impression of *Hallucigenia*. (e) A *Wiwaxia* fossil.

Most life existed on the sea bed — not floating like modern fish. The dominant life form was *trilobites*, creatures with a hard shell that have given a substantial fossil record (Figure 11.2a). They had segmented bodies and jointed limbs and also primitive eyes. Other creatures that existed were sponges and others that resembled starfish. A worm-like creature with fins, *Pikaia* (Figure 11.2b) had nerve fibres running along its length which connected the brain to various organs of the body — a construction that would eventually evolve into a backbone.

The most fearsome predator of the Cambrian period was the *Anomalocaris*, which could grow up to two metres in length. There are fossils of this creature, but the artist's impression in Figure 11.2c shows its form more clearly. It preyed on almost all the other creatures of the period — trilobites and various worms and molluscs.

The aptly named *Hallucigenia* (Figure 11.2d), walking on seven pairs of spiky legs and looking like the product of a bad dream, had two rows of spines on its back to protect it from attack from above. This sort of defence was also used by *Wiwaxia* (Figure 11.2e), resembling a rugby ball with spikes on its top surface. Surface dwellers, such as *Hallucigenia* and *Wiwaxia* were unprotected on their undersides. Trilobites developed a strategy of burrowing under the sea floor so that it could attack *Wiwaxia* and other smaller creatures from underneath where they were vulnerable.

There were four occasions during the Cambrian period when glaciations and mass extinctions occurred. When large amounts of water became ice, shallow seas would dry up and the life in them would die out. Again, creatures living in warmer regions could not adapt to colder conditions and became extinct. The Cambrian period ended with a major extinction event that wiped out a large proportion of the species that had developed.

By the end of the Cambrian period some sea creatures had moved onto land in damp regions adjacent to oceans, but there were no true flora on land, although there may have been algae and lichens in damp locations.

11.4.2 *The Ordovician period (488–444 Ma BP)*

Conditions were warmer at the beginning of the Ordovician period and many new life forms developed. Trilobites had survived the final Cambrian extinction and *cephalopods*, similar to the octopus, had evolved and become the major sea predator. There were also sponges, corals, *gastropods* — snail-like creatures — and *echinoderms*, which were related to starfish but had a feathery, fern-like appearance. Vertebrate fish were present in this period, possibly having evolved at the end of the Cambrian period. These fish had jaws; prior to this time fish had suckers, like lampreys, rather than mouths. A new and abundant species in this period were *brachiopods* (Figure 11.3), creatures like modern clams but biologically completely different.

Faunal life existed mainly in the sea, with some inhabiting tidal margins of land. During this period, the first flora, flat-branching, ribbon-like plants resembling liverwort, had moved onto land.

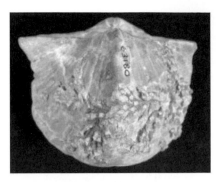

Figure 11.3   A brachiopod.

By the end of the Ordovician period, Gondwana was over the South Pole and large quantities of water were contained there in the form of a layer of ice several kilometres thick, as exists in Antarctica at present. The consequent fall in sea level drained shallow seas leading to the death of 70% of all living organisms by the end of the period.

11.4.3 *The Silurian period (444–416 Ma BP)*

Brachiopods, echinoderms and *nautiloids,* types of cephalopod, survived into the Silurian period. The climate was warmer, which melted the ice covering Gondwana and increased the depth and extent of the oceans. Collisions of tectonic plates created mountains and some previously submerged areas became dry land. Some plants that had adapted to tidal regions — wet but not always submerged — now adapted to drier conditions. These plants were very small but with rigid stems and root-like systems were the forerunners of present-day plants.

By this time, the ozone layer was well established, reducing the danger of damaging solar ultraviolet radiation. Tidal regions became nurseries where various kinds of fauna could adapt to living on land; arthropods — insects, spiders and centipedes — moved onto land, some of them much larger than present-day creatures of similar kinds. The fierce predator *eurypterid* (Figure 11.4), similar to a scorpion, was up to three metres in length and lived in shallow water since it could not support its full weight on land. In the sea, fish with backbones and jaws were becoming much more common.

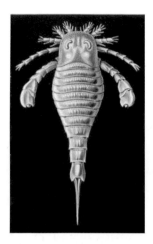

Figure 11.4   Eurypterid (illustration from Ernst Haecel, *Kunstformen der Natur*, 1904).

11.4.4 *The Devonian period (416–360 Ma BP)*

In this period, there was considerable mountain building due to the collision of two smaller continents, *Euramerica* and *Baltica*, to form Laurasia. The two super-continents, Laurasia and Gondwana, spanned the equator making conditions on land warm and conducive to the development of plant life. Plants developed true root systems that exploited moisture over a large area and vascular tissues to distribute it to all parts of the plant. Plants grew much bigger and propagated by producing seeds rather than spores as had the first land plants. The first true tree, *Archaeopteris* (Figure 11.5) had fern-like leaves and grew to a height of 15 metres or more — a tree that would not be much out of place in a modern setting. Some Devonian fauna and flora were developing forms that we would recognize today.

New species of fish were developing, such as the armoured *placoderms* (Figure 11.6) and sharks that probably evolved from placoderms by loss of the surface armour; both types of fish had cartilage backbones. Great reefs formed in the sea, similar to present-day reefs but derived from different living organisms. Many new invertebrate species, including ammonites (Figure 10.3) also evolved in the sea at this time.

Figure 11.5   An early true tree, *Archaeopteris.*

Figure 11.6   A typical placoderm (Tiere der Urwelt, *Creatures of the Primitive World*, 1902).

A particularly interesting Devonian fish is the *coelacanth* (Figure 11.7), thought to be long extinct. In 1938, a living coelacanth was caught in the Indian Ocean, since when others have been found from time to time.

Life was now well-established on land. Some fish had developed structured fins that had enabled them to 'walk' on the sea bed and these evolved into the legs of the first four-legged land-walking creatures (*tetrapods*), which were amphibian. One of these amphibians, *Ichthyostega* (Figure 11.8) clearly shows the evolution from a fish-like form.

Figure 11.7 An artist's impression of a coelacanth (Robbie Cade).

Figure 11.8 *Ichthyostega.*

11.4.5 *The Carboniferous period (360–299 Ma BP)*

During this period, Gondwana and Laurasia were coming together to form Pangaea (Figure 9.3). This produced mountains and the uplifting of land in some regions; Charles Darwin noted the presence of fossilized sea shells at high altitudes in South America and at this time a great deal of limestone (calcium carbonate, $CaCO_3$) was being produced on land from the shells of various sea creatures. The favourable damp and warm conditions led to an explosive increase in the number and types of plants. The residue of dead trees and ferns from this period is the source of present-day coal deposits.

Trilobites died out by the end of the period but brachiopods were still widespread. Tiny filter-feeding animals, a few millimetres in average dimension, called *bryozoa* (Figure 11.9), which attached themselves to rocky surfaces in great colonies and occupied the many shallow seas that existed at this time.

Placoderms became extinct and were replaced by many other species of modern-looking fish.

The large expansion of plant life and the resulting photosynthesis greatly increased the amount of atmospheric oxygen, which enabled

Figure 11.9   Forms of bryozoa (Ernst Haecel, *Kunstformen der Natur*, 1904).

Figure 11.10   An impression of an early reptile, *Hylonomus*.

the development of ever more complex creatures with higher rates of metabolism. Reptiles developed from amphibians and colonized the land well away from large stretches of water. Their leathery skins were better adapted to living under drier conditions; another important development was reproduction via an egg with a hard, originally leathery outer coating that prevented the contents from drying out. The greater likelihood of survival of the hatchlings meant that fewer eggs had to be laid. The early reptiles were small, typically 20 centimetres or so in length (Figure 11.10). By contrast, trees had become very large, growing up to 40 metres in height.

Insects also flourished in the warm, damp and well-oxygenated conditions. A giant precursor of the dragonfly, *Meganeura*, had a

Figure 11.11   The reptile *Dimetrodon* from the Permian period.

wingspan of up to 75 centimetres and there were giant millipedes, *Arthropleura*, that were 1.5 metres long.

11.4.6 *The Permian period (299–251 Ma BP)*

Pangaea had formed and was drifting northwards, giving cooler conditions. The interior of Pangaea had desert-like conditions with hot days and cold nights. Ferns and conifers were the dominant vegetation.

The drier conditions led to reptiles taking over from amphibians as the major land-dwellers; a common reptile of the time was *Dimetrodon* (Figure 11.11). Some reptiles took on mammalian characteristics and would evolve into mammals in due course — giving birth to live young (with few exceptions) and with females having mammary glands to feed infants with milk.

A great extinction of species of unknown cause occurred at the end of the Permian period. More than 90% of the marine species present became extinct; corals, which had thrived during the greater part of the Permian period, were decimated. Reptiles were most able to survive this extinction and they were destined to become the dominant species on land.

11.5 The Mesozoic Era (251–65.5 Ma BP)

This era began with reptiles as the dominant life form, but by the end of the era dinosaurs had become dominant. This era also saw the arrival of flowering plants, birds and mammals. It began with a great extinction and finished with another, and even greater, extinction of species.

(a)   (b)

Figure 11.12   Triassic period marine reptiles. (a) Ichthyosaur. (b) Plesiosaur.

11.5.1 *The Triassic period (251–200 Ma BP)*

During this period, Pangaea broke up into Gondwana and Laurasia and by the end of the period the continents were positioned as shown in Figure 9.10.

A common fossil of this period is that of the *belemnite*, an invertebrate like a squid but with an internal structure consisting of bony plates. The hip structures of some reptiles changed so that they adopted a more upright posture and reptiles also became dominant sea creatures, although still breathing through lungs rather than gills. Two examples of Triassic marine reptiles are *ichthyosaurs* (Figure 11.12a), some eight metres in length, and *plesiosaurs* (Figure 11.12b), about 12 metres long, including its long neck; the fabled monster 'Nessie', reputed to live in Loch Ness in Scotland, is claimed to resemble a plesiosaur. Towards the end of the Triassic period, there were some flying reptiles, *pterosaurs* (Figure 11.13), but it is not thought that they were the ancestors of modern birds.

The first dinosaurs appeared towards the end of the Triassic period. Although the Greek word σαυρα (saura) means lizard, dinosaurs are different from lizards in three important respects. Firstly, they were warm-blooded, unlike the cold-blooded lizards, secondly, because of their hip and leg structures they could stand upright and thirdly, they had hands that could be used for grasping objects. Dinosaurs would eventually rule the Earth — but not yet.

Figure 11.13   An artist's drawing of a pterosaur.

Figure 11.14   The mammal-like lizard *Thrinaxodon*.

Another significant development towards the end of the period was the appearance of reptiles that resembled mammals, for example *Thrinaxodon* (Figure 11.14), a carnivore about 30 centimetres long, which may have been fur-covered and warm-blooded. True mammals later evolved from creatures of this kind. However, cold-blooded lizards were better able to survive the heat of Triassic deserts and the mammal-like lizards may have thrived mainly in the cooler regions of Earth at that time.

The trees and ferns that had flourished in the Carboniferous period no longer survived in the comparatively dry Triassic period. The dominant vegetation was now conifers and *Ginkgos*, the latter being giant trees that are still found in some parts of eastern Asia.

11.5.2 *The Jurassic period (200–145 Ma BP)*

Laurasia and Gondwana were breaking up to form the continents that exist today and there was abundant volcanic activity. Plate collisions produced mountains, especially along the western side of Pangaea, corresponding to the western coasts of present-day North and South

(a) (b)

Figure 11.15   Herbivorous dinosaurs of the Jurrasic period. (a) *Brachiosaurus*. (b) *Stegosaurus*.

America. The climate was warmer than now and the dominant plant life consisted of palm-like trees, conifers, ferns and various smaller species. The warm and stable climate over the period allowed the development of many new species of plants and animals.

Ichthyosaurs and plesiosaurs increased in number, as did belemnites and ammonites, all of which were efficient predators. There was an abundance of vertebrate fish, sharks and rays.

Dinosaurs dominated the animal world. The herbivores had an abundance of food and grew to huge sizes. The largest dinosaurs of the period, *sauropods*, stood on four legs, had long necks and a counterbalancing long tail; a typical sauropod, *Brachiosaurus* (Figure 11.15a), weighed up to 55 tonnes. Another type of herbivorous dinosaur was the *ornithopod*, which had hip structures resembling that of present-day birds. Figure 11.15b shows *Stegosaurus*, about nine metres long with two rows of spines along its back. It did not have a long neck so it almost certainly fed from ferns and other plants close to the ground.

The carnivorous dinosaurs of this period were also gigantic — a typical example being *Allosaurus* (Figure 10.6).

Many true mammals had evolved, although they were small mouse-like creatures. Some dinosaurs, such as *Archaeopteryx* (Figure 10.5), were taking to the air; they are believed to be the ancestors of modern birds.

Like many periods, the end of the Jurassic period is marked by a minor mass extinction.

(a) (b) (c)

Figure 11.16 Three dinosaurs of the Cretaceous period. (a) *Triceratops.* (b) *Hadrosaurus.* (c) *Tyrannosaurus rex.*

11.5.3 *The Cretaceous period (145–65.5 Ma BP)*

The Atlantic Ocean was beginning to form by the break-up of Laurasia and Gondwana with the ocean rift producing many undersea volcanoes and mountains. The sea level was much higher than now, by as much as 100 metres.

Flowering plants and new kinds of grasses and trees developed. The flowers developed a symbiotic relationship with insects that benefited from the nectar produced by the flowers while they performed the task of efficiently moving pollen from one flower to another.

New dinosaurs that appeared in this period included *Triceratops*, a large bird-hipped herbivore with three horns and a bony collar (Figure 11.16a), the herbivore *Hadrosaurus* (Figure 11.16b), up to ten metres long and weighing up to seven tonnes, and the fearsome carnivore *Tyrannosaurus rex* (Figure 11.16c), which was up to twelve metres long, five metres tall and was the largest carnivorous creature to ever walk the Earth.

Mammals were also on the increase in this period, but they were an insignificant part of the fauna. A new feature of mammals was the evolution of the placenta and the development of offspring within the female's body.

The KT event (Cretaceous-Tertiary — the K in KT is from the German name for this period — Kreidezeit, Section 10.3) ended the Cretaceous period and also heralded the beginning of the Tertiary period, the first period of the next era.

While dinosaurs dominated the Earth, there was no opportunity for any other type of creature to evolve to a size that could challenge them.

Early mammals had developed, but their defence was that they were small and inconspicuous and that they did not compete with dinosaurs in any direct way. Now that dinosaurs had vanished, the field was clear for mammals to evolve and to become dominant.

11.6 The Cenozoic Era (65.5 Ma BP to Present)

This era is divided into two periods, the *Tertiary* and *Quaternary*, which are each further subdivided into shorter epochs that enable us to follow evolutionary processes in greater detail as we approach the present.

11.6.1 *The Tertiary period (65.5–1.8 Ma BP)*

In this period, mammals became the dominant form of fauna. Early in the period, the first primates appeared — the first step on the evolutionary ladder towards modern man. These early primates had some human characteristics — five digits on their fore and hind limbs with opposable thumbs, thus enabling them to grasp and manipulate objects, and eyes with good colour vision located at the front of the face, giving them good stereoscopic vision. Some primates could hold their bodies upright and walk on two legs — which freed their hands for other activities while they walked. Another common characteristic is that, compared to other creatures, they had a large brain relative to their size.

The period ends with the evolution of hominids, mammals that are definitely man-like in structure and appearance, and possibly the first predecessor of modern man had appeared on the scene.

11.6.1.1 *The Palaeocene epoch (65.5–56 Ma BP)*

Europe and North America were partially joined, Australia was attached to Antarctica and India was still on its journey towards Asia.

The only remnants of the dinosaurs were birds that, by their nature as flying creatures, could not be physically very large. Mammals, both herbivores and carnivores, were increasing in size, although still quite small. The mammals of this period had not yet developed the specializations that make modern mammals so successful in survival

Figure 11.17 *Hyracotherium.*

terms. Some of them still laid eggs, like the modern platypus, and many were marsupials, with pouches like kangaroos.

Broad-leafed trees covered large parts of the Earth and provided a habitat for many early mammals. There is fossil evidence that a squirrel-like arboreal creature which may have been a precursor of primates evolved during this time.

11.6.1.2 *The Eocene epoch (56–34 Ma BP)*

Now the Atlantic Ocean had completely formed, from pole to pole. At the beginning of the period the temperature was so high that tropical rainforests existed at the poles but by the end of the period the Antarctic icecap had formed and has been there ever since.

The *Hyracotherium* (Figure 11.17), the forebear of horses and similar creatures, appeared on the scene, although they were small — about the size of an average dog. Horses, zebras and the rhinoceros are the modern animals from this branch of evolution.

Some mammals reverted from living on land to living in the sea. The ancestors of whales appeared, although they were unlike the whales we know today. A stage in this evolutionary process is *Ambulocetus* (Figure 11.18), a mammal about three metres long. Like a modern whale, it had no external ears and had an adaptation of its nose that enabled it to swallow under water. It was a carnivore that probably hunted much as a crocodile hunts today, by hiding under water and grabbing at its prey.

Figure 11.18 *Ambulocetus,* an ancestor of the whale.

Figure 11.19 Lemurs.

The first primates appeared, resembling modern *prosimians* such as tarsiers, bush babies and lemurs (Figure 11.19). A prosimian is a primate that is neither a monkey nor an ape. For a primate to take an upright posture the hole in the skull where the spinal cord enters the head, the *foramen magnum,* must move from the back of the skull towards the centre. This was happening during the Eocene epoch, suggesting that, like modern lemurs, when sitting or moving about the prosimians of the time had their bodies in an erect position.

During this epoch, prosimians were present everywhere that had forests and a suitably high temperature. However, by the end of the epoch the first monkeys appeared and most species of prosimians became extinct because they could not compete with the new arrivals. Only on the island of Madagascar, where monkeys and apes did not evolve, were the prosimians able to flourish, and lemurs still exist there in great numbers. In other locations lemurs have become nocturnal creatures, keeping them clear of the larger primates.

Figure 11.20 A drawing of a *Paraceratherium*, compared to a man.

11.6.1.3 *The Oligocene epoch (34–23 Ma BP)*

Within this period, the Indian plate crashed into the southern flank of Laurasia, and Antarctica, which had separated from Australia, reached the South Pole and was becoming ice covered. The world climate changed from very wet and tropical to somewhat drier and sub-tropical and grasses developed that produced huge savannah regions.

The change in conditions favoured the domination of mammals such as deer, horses, cats and dogs and the first elephants appeared. The largest mammal from this epoch was *Paraceratherium*, a giant hornless precursor of the rhinoceros, which was more than six metres tall (Figure 11.20). The formation of thick Antarctic ice sheets caused the seas to retreat and there was a loss of many marine species, including the first primitive whales, which were later replaced by modern whales.

The prosimians evolved into the first *Old World monkeys*, which occupied tropical and grassland regions of Africa and Asia. They were mostly small, although some approached the size of a small dog. A type of Old World monkey is the *talapoin* shown in Figure 11.21. It is easy to imagine a slow evolutionary change from a lemur to this creature; the snout has become smaller, it has fewer teeth, eyes more forward looking and a larger brain. It weighs about one kilogram and is tree dwelling. At the other extremity of Old World monkeys the modern *mandrill* lives on the ground and can weigh up to 50 kilograms.

Figure 11.21   Talapoin, an Old World monkey (Friedrich Wilhelm Kuhnert).

In the early part of the epoch, *New World monkeys* appeared in South America. These are thought to have derived from earlier Old World monkeys in Africa that migrated across the Atlantic Ocean on floating islands of vegetation. However, by whatever means they reached South America, they evolved into new distinctive forms. They have flatter noses, with nostrils at the side, and they have long prehensile tails, useful for holding on to branches — something the shorter-tailed Old World monkeys cannot do. Another interesting difference is that all the New World monkeys, with the exception of *howler monkeys*, lack full-colour vision. They have only two visual pigments in their retinal photoreceptors, rather than the three for Old World monkeys, and so they have a more restricted range of colour discrimination.

11.6.1.4 *The Miocene epoch (23–5.3 Ma BP)*

The arrangement of land was similar to that at present, except that North and South America were still separate. The Arctic icecap had formed, but, since the ice was floating, this had no implications for sea levels. Mountain ranges were building in the western Americas and northern India. The climate was cooler and drier and extensive grasslands provided a good environment for herbivores. Most fauna consisted of modern species such as wolves, horses, camels, deer, crows, ducks, otters and whales.

The first apes evolved from monkeys, distinguished from them by not having tails, by usually being larger and by having a more human appearance. The family of *Hominoidea* or *great apes* include chimpanzees, orang-utans, gorillas, bonobos (a dwarf chimpanzee) and

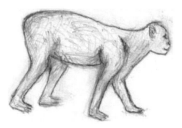

Figure 11.22 An early ape — *Proconsul.*

humans. Like their monkey forebears, they are mostly tree-dwelling — the exceptions being the gorilla and humans. There is also a family of *lesser apes*, which includes gibbons of various kinds. One of the early apes was the *Proconsul* (Figure 11.22), a tree-living ape that occupied the forests of Africa between 21–14 Ma BP.

Altogether about 100 different kinds of ape evolved, one or more of which were precursors of *hominids*, the ancestors of modern man. They lived mainly in Africa and in southern grassland area of Europe. However, colder conditions in Europe towards the end of the Miocene epoch led to the extinction of many species of apes; the remainder survived in the more benign and warmer conditions in Africa and by migration to Southern Asia.

11.6.1.5 *The Pliocene epoch (5.3–1.8 Ma BP)*

The land bridge formed between North and South America, which allowed the mixing of the species in the two continents. Ice sheets covered both poles and the Earth steadily became cooler.

Primates continued to evolve, with hominids, close relatives of man, having prominent jaws and larger brains, appearing towards the end of this epoch. The oldest fossil record of a hominid dates from just over five million years ago and was discovered in Ethiopia. A younger, almost complete skeleton of a female hominid, called Lucy by its discoverers and designated as of species *Australopithecus africanus*, dates from about 3 Ma BP and may represent the stage at which chimpanzees and humans diverged in their evolution. It is notable that humans and chimpanzees have 98.9% of their coding-DNA in common. Lucy was 1.1 metres tall and weighed about 29 kilograms. From her pelvic

Figure 11.23   *Australopithecus afarensis* (Cosmo Caixa, Barcelona).

bones it can be deduced that she walked upright on two legs. An artist's impression of the hominid, *Australopithecus afarensis*, earlier than Lucy, as deduced from its fossil remains, is shown in Figure 11.23.

An early hominid that appears at the end of the Pliocene epoch, about 2.2 million years BP, which recognizably spans the appearance gap between *Australopithecus africanus* and modern man is *Homo habilis* (Figure 11.24a). *Homo habilis* had a brain size in the range 500–800 cubic centimetres, greater than that of *Australopithecus africanus*, which was about 450 cubic centimetres, and a part of the brain associated with speech had developed somewhat, so he may have been capable of elementary speech communication. There is evidence that he was making and using primitive stone tools. He was of similar size to *Australopithecus africanus*, with average height 1.5 metres and weight 45 kilograms — which is also similar to that of modern pygmy people in the Congo region of Africa.

The end of the Tertiary period marks the beginning of the evolutionary process that leads to modern man — *Homo sapiens* — all happening in a space of about three million years. Seeing the changes

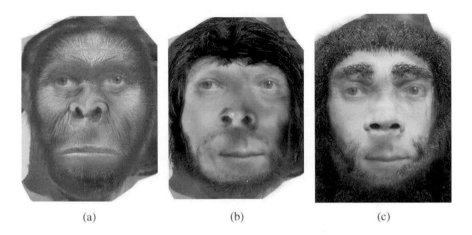

(a)   (b)   (c)

Figure 11.24   (a) *Homo habilis.* (b) *Homo erectus.* (c) *Homo heidelbergensis.*

in that period from the creature portrayed in Figure 11.23, probably not very different from Lucy, to man as seen today gives a better understanding of how changes wrought over 4,000 million years could give present life forms, starting with the simple bacterium.

11.6.2 *The Quaternary period (1.8 Ma BP to present)*

This is a period of very rapid change, in particular in the rate of development of the capability of the species that evolved, by which we mean hominid development leading to mankind. From *Australopithecus africanus* various strands of development led towards ever more advanced forms of hominid in terms of brain size and concomitant ability to manipulate nature to its own ends. Many of these evolutionary strands, separated by migration into different environments, evolved separately to become distinctive in both appearance and lifestyle. Most of these strands became extinct, although their fossil remains are frequently uncovered. Thus, *Homo antecessor*, the fossil remains of which were discovered in Spain in 1977, is dated to 780,000 years ago and shows a mixture of characteristics of modern man and of more primitive hominids. There is even doubt that the fossil remains are all of one species but, in any case, it is difficult to link *Homo antecessor* into the line of development of modern man. Here we shall concentrate on three kinds of hominid that seem to indicate a sequence of development between *Homo habilis*, the first primate distinguished by the term 'homo', indicating 'man', and modern man, described as *Homo sapiens* — 'wise man'.

11.6.2.1 *The Pleistocene epoch (1.8 Ma–11,500 years BP)*

The climate of this period was punctuated by repeated glacial cycles that, at their peak, covered 30% of the Earth's surface. At maximum glaciation a great deal of the Earth's water was in the form of ice and the seas retreated. In warmer periods the seas advanced again. Because of the decrease in liquid water available to be evaporated, the climate became drier and extensive deserts formed.

Species of animals that became extinct by the end of this period include mammoths, mastodons (like woolly elephants), sabre-toothed tigers, *Glyptodons* (similar to armadillos and as big as a small car) and ground sloths.

At the beginning of the epoch, about 1.8 Ma BP, a new kind of hominid, *Homo erectus,* appeared on the scene. From fossil skulls it is deduced that he was beginning to resemble modern man (Figure 11.24b). His brain size was between 750 and 1,225 cubic centimetres, a considerable advance on *Homo habilis;* his stone tools were much more sophisticated than those of *Homo habilis* and there is some evidence that he had mastered the use of fire. *Homo erectus* is found in many locations in the world, in Africa, Asia and Europe, with slightly different physical characteristics in different locations. The overall impression from the various fossil remains is that they were of similar stature to modern man but much more robust.

The next stage in the evolutionary pathway to modern man is an early form of *Homo sapiens, Homo heidelbergensis* (Figure 11.24c), which was certainly present 500,000 years ago but may have been slightly earlier. This species of hominid bridges the gap between *Homo erectus* and modern man. The brain size averaged 1,200 cubic centimetres but was larger for some individuals, and the skull shape was closer to that of modern man, although still with prominent brow ridges and a receding forehead and chin. They were skilful stone-tool makers and had mastered the use of fire.

Humans essentially attained their present form during this epoch. There were two lines of human development, *Neanderthals*, who developed in Europe and western Asia about 230,000 years ago, and *Homo sapiens*, who first evolved in Africa. Neanderthals were shorter and more heavily built than *Homo sapiens* and were well adapted to living in a cold climate. They appear to have had larger brains, although, since they were heavier than modern man, the ratio of brain size to weight was probably similar. When the skull and other bones of Neanderthal man were first discovered in the early 19th century, images of their postulated appearance were such that they would not have seemed out of place in a zoo. While the shape of their skulls was different from that of modern man — they had a distinct bulge at the back and were distinctly beetle-browed and had flatter noses that suited living in a cold climate — the dominant view now is that, while they would certainly have stood out in a modern setting, they were mostly similar in facial appearance to present man. An impression based on this interpretation of their appearance, is given in Figure 11.25. Neanderthals coexisted with *Homo sapiens* for some time and the cause of their extinction is not fully understood. The last traces of them were found at Gibraltar, where they existed until 24,000 years ago.

11.6.2.2 *The Holocene epoch (11,500 years BP to present)*

This is the age of modern man. The climate has been relatively stable and warm during this period with minor blips, such as the few hundred years of lower temperatures ending in about AD 1800. During the reign of Elizabeth I, the river Thames occasionally froze over to the

Figure 11.25   An impression of Neanderthal man.

extent that the people would hold a 'frost fair' on the Thames itself. Such an event is now unimaginable.

Homo sapiens has spread over the whole habitable globe in exponentially increasing numbers and has developed technology to the point where the environment is as much influenced by man as by natural events — with results as yet uncertain but a cause for concern.

Chapter 12

The Ages
of Archaeological Remains

12.1 Archaeology and Palaeontology

Archaeology is the study of the history of mankind through the dating and understanding of the products of different societies through the ages. These can include artefacts of various kinds, from primitive tools to the complex products of the Industrial Revolution, buildings from primitive shelters to massive constructions such as Egyptian pyramids and human remains, which can often reveal the living conditions of the time. The dating of remains and objects from the period before the arrival of man is within the field of *palaeontology*.

In a sense, we covered some aspects of archaeology in Chapter 11, where the progress from primitive hominids to *Homo sapiens* was dealt with as part of the development of life on Earth. Determining the ages of the remains of such beings was mostly done in the same way as with other creatures but for more recent hominid finds, such as those of Neanderthals from 24,000 years ago, the techniques described in this chapter are also appropriate.

Most archaeological studies are concerned with the study of human societies and civilizations going back about 9,000 years — approximately within the Holocene epoch. Although the study of the development of human societies is fascinating, here we shall just be concerned with techniques of dating objects, which may be applied equally to the remains of a simple society in the Orkney Islands and to those of the contemporary complex civilization of ancient Babylon.

12.2 The Occurrence of Carbon-14 on Earth

In Chapter 8, we described the techniques used in radioactive dating. Here we describe the dating of archaeological remains using the radioisotope of carbon, C-14, the most widely used and precise dating tool in the armoury of the archaeologist.

Carbon dioxide constitutes 0.035% of the atmosphere with isotopic composition for carbon 89.9% C-12, 1.1% C-13 and C-14, an unstable isotope with a half-life of 5,730 years present at a level such that the [C-14]/[C-12] ratio is about 1.3×10^{-12}. The decay process involves the emission of an electron to give the common stable isotope of nitrogen, that is

$$C\text{-}14 \rightarrow N\text{-}14 + \text{electron}. \qquad (12.1)$$

There is actually another particle emitted, called an *electron antineutrino*, which has no charge and a tiny mass but, while it is an important particle for the study of particle physics, it can be ignored for our present purpose. Since C-14 is radioactive with a half-life that is short by geological standards, it might be wondered how any of it exists in the atmosphere. The answer is that it is constantly being created in the lower reaches of the stratosphere at heights between about 10 and 15 kilometres by the action of cosmic rays which, despite their name, are actually mostly very energetic particles of various kinds. The reaction that produces the C-14 involves the interaction of a neutron with the common stable isotope of nitrogen, N-14, which is the majority component — about 78% — of the atmosphere. This reaction is

$$N\text{-}14 + \text{neutron} \rightarrow C\text{-}14 + \text{proton}. \qquad (12.2)$$

The balance of the rate of production and the rate of decay maintains the tiny component of C-14 in atmospheric carbon dioxide.

All advanced vegetable and animal life is ultimately dependent on atmospheric carbon dioxide; too much of it may produce global warming but if it were not present in the atmosphere, then life as we know it could not exist — for one thing the average temperature of the Earth would be 33°C lower than it is now. The process of photosynthesis in plants — which we have already referred to in relation to cyanobacteria — converts carbon dioxide plus water into cellulose plus oxygen. The cellulose constitutes the main structural element of most plants and the oxygen is the essential component that fuels the metabolic processes upon which most life depends. Carbon dioxide allows plants to grow, herbivores and omnivores consume plants and carnivores and omnivores consume herbivores and omnivores.

The net effect of this food chain, which begins with atmospheric carbon dioxide, is that all the carbon in living organisms contains the same proportion of C-14 that exists in the atmosphere. However, once the organism dies, it ceases to be in a state of equilibrium with atmospheric carbon dioxide and the proportion of C-14 in the carbon residue of the decaying or decayed organism will decrease with time. It is detecting this decrease that is the basis of C-14 dating.

12.3 The Techniques of Radiocarbon Dating

The idea of using C-14 to date carbon-containing materials from the past, known as *radiocarbon dating*, was proposed in 1949 by the American physical chemist Willard Frank Libby (1908–1980), an idea that earned him the Nobel Prize for chemistry in 1960. Radiocarbon techniques can successfully date materials back to about 50,000 years BP, a period which takes in the whole period of the development of ancient civilizations. The kinds of material that can be dated include wood, leather, bone, shells and textiles.

The basic equation describing the decay, which is similar to equation (8.1a), is

$$[C\text{-}14]_t = [C\text{-}14]_0 \exp(-\lambda t), \qquad (12.3)$$

where $[C\text{-}14]_0$ is the initial concentration of C-14, $[C\text{-}14]_t$ is the concentration after time t and λ is the decay constant, which is 1.2097×10^{-4} year^{-1}, corresponding to a half-life of 5,730 years; with this form of λ, t in equation (12.3) must be expressed in years.

There are two main methods for radiocarbon dating, both dependent on equation (12.3). In our description of these methods we shall firstly assume that the initial ratio of carbon isotopes, $[C\text{-}14]_0/[C\text{-}12]$, is invariant over time, i.e. the same at the times of ancient Egypt and the reign of Elizabeth I of England as it is now. We shall see that this assumption is false, and a correction in the determined age must be made for the variability of the ratio.

Willard Frank Libby

12.3.1 *Removing contamination and preparation of samples*

An essential requirement to be able to obtain the radiocarbon age of an object must be to remove any source of contamination containing carbon. There are two extreme cases. In the first, the contaminant may be either living material, such as a plant root that has invaded a wooden artefact, or non-living material, such as the paper in which the object was wrapped, which is derived from recently-living material. In such a case the [C-14]/[C-12] ratio is raised above that for the uncontaminated object and the apparent age will be lowered. Figure 12.1 shows the indicated age against true age for 1% and 5% of modern-carbon contamination.

In the second extreme case, the contamination can be with carbon from an ancient source, which could be a mineral or an extremely old residue of living matter, both of which will contain no C-14. In this case the [C-14]/[C-12] ratio will be lowered and the object will have an indicated age more than the true age. Figure 12.2 shows the effect of 1% and 5% contamination with old carbon; it will be seen that the overall effect is less than for contamination with modern carbon and is virtually constant over the whole age range.

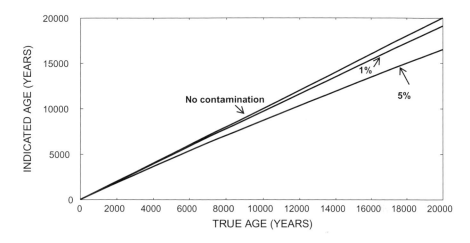

Figure 12.1 Indicated age against true age for 1% and 5% contamination with modern carbon.

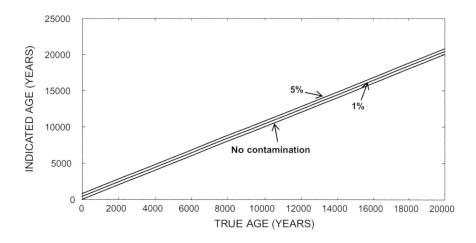

Figure 12.2 Indicated age against true age for 1% and 5% contamination with ancient carbon.

The way that contaminants are removed depends on the type of object and the environment in which it is found. Scientists in this field have a considerable arsenal of ways to deal with contamination, built up over decades of experience, for example, objects of wood, charcoal or textiles are usually washed with hot hydrochloric acid, followed by a wash in sodium hydroxide (caustic soda) and finally another wash in hydrochloric acid. If the object is contaminated with plant matter, then this can be physically removed or an object with only surface layers contaminated can have the outside layers scraped away.

Once the contamination has been removed, the next process is to extract from the specimen to be dated a sample of pure carbon. This can involve many steps, which must all involve processes that cannot add carbon to the specimen. The way that Libby prepared his specimens gives an idea of the complexity of the processes used. Having removed any contaminants, he first converted the carbon in the specimen to carbon dioxide; for organic specimens this was done by combustion and for shells (calcium carbonate) by addition of hydrochloric acid. The gas was then passed through hot copper oxide that oxidized any carbon monoxide in the gas to carbon dioxide. After drying, the carbon dioxide was converted to carbon by passing it over heated magnesium. The resultant mixture of magnesium, magnesium oxide and carbon was then treated with hot hydrochloric acid which dissolved the magnesium and magnesium oxide but had no effect on the carbon. Straining off all the liquid products and washing with distilled water gave a sample of pure carbon that could be used to determine the age of the specimen.

12.4 The Beta-Counting Method

In this method the age of the specimen is estimated from the rate at which β-particles are emitted, since the rate of emission is proportional to the amount of C-14 present, which is an indicator of age. There are two main methods of beta-counting, either using a *gas proportional counter* or a *liquid scintillation counter*, the former method having been largely, but not completely, superseded by the latter.

The main element of a gas proportional counter is a chamber containing an inert gas, which will produce ions when exposed to ionizing radiation or energetic particles, plus a *quench gas* that will quickly terminate the ionization process so that the chamber is ready to detect the next ionization event; the counter will only be proportional if the quenching time is appreciably shorter than the expected interval between ionizing events. A very common combination of gases is 90%

argon and 10% methane. By converting the specimen carbon into methane, it can act simultaneously as the source of the emissions to be recorded and as a quench gas.

A β-particle emitted by a C-14 atom has energy 0.15 MeV.[1] When it collides with an argon atom, the atom is ionized to give an electron plus an argon ion. This ion then ionizes other argon atoms as it moves through the gas, with the total number of ion pairs produced being proportional to the energy of the original β-particle; about 5,000 ion pairs will be produced by a C-14 β-particle. An electric field between electrodes at opposite ends of the chamber causes the negatively-charged electrons and positively-charged ions to separate and the effect is that an electric pulse is recorded, the energy of which is proportional to the energy of the original ionizing particle. Since the energy of the C-14 β-particle is known, any spurious pulses due to particles or photons coming from outside the chamber can be discounted — although the apparatus is heavily shielded to reduce such spurious events to a minimum.

In a liquid scintillation counter the carbon is converted into benzene, C_6H_6, a liquid, and a scintillating material is dissolved in the benzene. A scintillation counter is shown schematically in Figure 12.3.

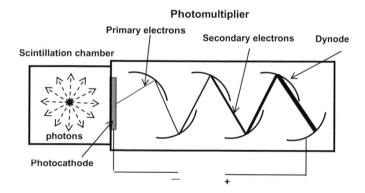

Figure 12.3 A schematic of a scintillation counter apparatus for radiocarbon dating. The photomultiplier will usually contain more dynodes.

---

[1] An electron volt (eV) is 1.602×10^{-19} joules (J). 1 MeV is one million electron volts.

The benzene, containing C-14 plus the scintillator, is situated in the scintillator chamber. Whenever a β-particle is emitted, it excites a number of scintillator atoms which then fall back to their ground state with the emission of photons of ultraviolet radiation (dashed arrows). Some of these impinge on the photocathode that emits *primary electrons*. A number of these move towards the first *dynode*, which is at a potential of 90–100 volts higher than the photocathode. The accelerated primary electrons strike a layer of material such as magnesium oxide (MgO) which coats the dynode and each primary electron produces about ten new electrons. These in their turn move on to the next dynode, again at a higher voltage than the previous one and, once again, the number of emitted electrons will increase by a factor of ten. If there are five dynodes, then by the time the electrons leave the final dynode their number will have increased by a factor 10^5 and the electrical pulse they deliver is recorded. This arrangement is called a *photomultiplier* and is used in many kinds of experiment. The total time from the original scintillation to recording the electrical signal is a few nanoseconds, so for scintillations at a rate less than 10^6–10^7 per second the counting rate will almost precisely equal the scintillation rate.

As a test of a beta-counting equipment, it is run with no sample present and also with a standard sample of known age. This provides a test that the apparatus is working properly and not producing spurious counts.

We now determine the order of magnitude of the counting rates that would be expected for radiocarbon dating by the β-counting method.

12.4.1 *Numerical aspects of beta-counting*

The objective in the beta-counting approach is to have a total number of counts that will give a reasonable precision in assessing age. If the number of emissions in a particular time interval T is N, that does not mean that there will be N emissions for every interval T. There will be a statistical fluctuation in the number; if counts were taken over a very large (theoretically infinite) number of intervals T, then the average of the counts would be the 'true count' for that interval. For a single count N the probability of its variation away from the true count is expressed in terms of the *standard deviation* σ which is $\sqrt{N}$ in this case. There is a 68% probability that the magnitude of the deviation from the true number is less than σ, a 95% probability that it is less than 2σ and a 99.5% probability that it is less than 3σ. For example, if $N = 10{,}000$, then σ is 100, just 1% of N, and if $N = 100$, σ would be 10% of N, giving a not very precise estimate.

The quantity of the sample material that is required for a beta-counting measurement depends on its carbon content. For charcoal, which is almost all carbon, about 20 grams is sufficient. For wood, where only the cellulose component is used for carbon extraction, about 100 grams is needed and for bone, which has comparatively little carbon content, about 400 grams. We will first estimate the counting rate for 20 grams of modern carbon.

Most carbon is C-12 and contains 12 nucleons, each with average mass 1.66×10^{-27} kg. Hence the total number of carbon atoms in 20 grams (0.02 kg) of carbon is

$$n = \frac{0.02}{12 \times 1.66 \times 10^{-27}} = 1.00 \times 10^{24}. \tag{12.4}$$

With $[\text{C-14}]/[\text{C-12}] = 1.3 \times 10^{-12}$, the number of C-14 is 1.3×10^{12}. The relationship for the variation of the number of C-14 atoms with time is, following equation (12.3),

$$n_{14}(t) = n_{14}(0) \exp(-\lambda_{14} t). \tag{12.5}$$

To find the rate of change of C-14 atoms due to β-emission, we must differentiate (12.5) with respect to time, giving

$$\frac{dn_{14}(t)}{dt} = -\lambda_{14} n_{14}(0) \exp(-\lambda_{14} t). \tag{12.6}$$

Thus, the number of counts per unit time for modern carbon, obtained by putting $t = 0$ in (12.6), is $\lambda_{14} n_{14}(0)$. From (8.2), expressing λ_{14} in units s^{-1} with the year as 3.156×10^7 s,

$$\lambda_{14} = \frac{0.6931}{\tau_{1/2}} = \frac{0.6931}{5730 \times 3.156 \times 10^7} = 3.83 \times 10^{-12} \, s^{-1}.$$

$$\tag{12.7}$$

Hence the rate of beta-emission for 20 grams of modern carbon is

$$r(0) = \lambda_{14} n_{14}(0) = 3.83 \times 10^{-12} \times 1.30 \times 10^{12} = 5.0 \, \text{s}^{-1}. \quad (12.8)$$

If the count were carried out for an hour, the number of counts would be 18,000 giving a standard deviation of 134, about 0.75% of the actual count. A longer measuring time would give a smaller standard error as a percentage of the actual count.

After one half-life the count rate would be down to 2.5 counts s^{-1} and after m half-lives it would be $5/2^m \text{s}^{-1}$. After eight half-lives, 58,730 years, the count rate would be 0.02 s^{-1} or 72 per hour and a very long counting time would be required to reduce the standard deviation to a reasonably low percentage of the actual total count. Beyond eight half-lives beta-counting is too imprecise to give reliable results.

12.4.2 *Sources of error in beta-counting*

There are three main sources of error in radiocarbon dating by beta-emission, assuming that contamination of the samples has been eliminated. These are:

(a) Experimental error — deficiencies in the equipment being used.
(b) Background radiation or particles giving spurious counts.
(c) Assuming that the [C-14]/[C-12] ratio is invariant with time.

(a) *Experimental error*

The care with which the measuring apparatus is designed and set up is of importance and in the best measuring laboratories this source of error will be insignificant by comparison with the other two, so we shall not consider it further. Another source of experimental error is the way in which the specimens for beta-counting are prepared and handled. For example, if a benzene sample is allowed to evaporate then the proportional loss of C-12, the lightest isotope, will be greater than that of C-13, which in its turn will be greater than that of C-14. This consequent increase in the [C-14]/[C-12] ratio will lead to an age estimate that is too young — by about 80 years if 1% of the benzene evaporates.

(b) *Background radiation*

It is necessary to shield the equipment by placing it within an absorbing metallic container, often made of lead, known as a *passive shield*, to remove as far as possible any external influences that would give spurious counts. However, cosmic rays are very penetrating and can pass through quite thick shields. An alternative or auxiliary technique is to use what is known as *active shielding* where the beta-counting apparatus and its passive shield is surrounded by a scintillation counter that intercepts any particle or radiation entering from outside. If a signal from the scintillation counter is coincident with one from the beta-counter, then it is assumed to be a spurious event and the electronic control system eliminates it from the beta-count.

(c) *Variation of* [C-14]/[C-12] *ratio*

The assumption that the [C-12]/[C-14] ratio is constant over the whole applicable period for C-14 dating techniques turns out to be invalid. A fundamental element in understanding variations in the ratio is the fact that the atmosphere contains only 1.9% of the total carbon that can potentially enter the atmosphere, the bulk of the remainder being dissolved in the oceans in the form of carbon dioxide. The carbon contained in surface ocean water is older than that in the atmosphere and the [C-14]/[C-12] ratio is about 0.95 that of atmospheric carbon. There is a constant exchange of carbon dioxide between surface ocean water and the atmosphere; when temperature increases some carbon dioxide is released from the oceans and, conversely, when the temperature falls more carbon dioxide is absorbed. There is also an exchange between surface water and deep ocean water with carbon in deep ocean water being older than atmospheric carbon by about 1,000 years and therefore depleted in C-14. It is a complicated system but the main point here is that there is an exchange between atmospheric carbon and a much larger reservoir of older carbon.

An important variation in the C-14 content of the atmosphere was made in the latter half of the 20th century by the testing of nuclear weapons which began in the 1950s and which within ten years had almost doubled the quantity of C-14 in the atmosphere. Mixing with reservoir carbon has now reduced the effect but even so, the level of C-14 in the atmosphere is still 10% or so higher than it was in the pre-bomb era. It would, of course, be wrong to use the beta-emission of modern carbon as the baseline level for the dating of artefacts, so the standard used as the zero-time [C-14]/[C-12] ratio is that in a stock of oxalic acid, $H_2C_2O_4$, kept by the US National Institute of Standards and Technology with a ratio equivalent to that of 1950 wood — from before the testing of nuclear weapons.

Another source of variation is the burning of fossil fuels, which began on a large scale at the beginning of the Industrial Revolution at the end of the 18th century. Fossil fuels, laid down in the Carboniferous period, have zero C-14 content, so while they have greatly increased the overall carbon dioxide content of the atmosphere, they added nothing to the inventory of C-14. However, once again, the mixing of atmospheric carbon with that in the ocean reservoir has diminished the effect to a reduction of about 0.2% in the [C-14]/[C-12] ratio.

What we have dealt with so far are influences on the C-14 content of the atmosphere due to human activity in the recent past. However, it turns out that over the period of interest for radiocarbon dating there have also been variations due to natural causes. There are possible variations in the intensity of cosmic rays interacting with the atmosphere, mainly due to variations in the Earth's magnetic field, with which the cosmic rays interact. This is likely to be a very minor effect over the time period of interest. Of greater significance is the variation in the mean temperature of the Earth, with glacial periods leading to a large scale production of ice. This ice would not be part of the carbon reservoir but by the time it melted its carbon dioxide content would have been depleted in C-14 and would have contributed mainly to surface ocean water that is most intimately in contact with the atmosphere. The consequent variation in the [C-14]/[C-12] ratio can be quite significant in terms of determining the ages of objects, so it is important to determine how the ratio has varied with time. This will be dealt with in Section 12.6.

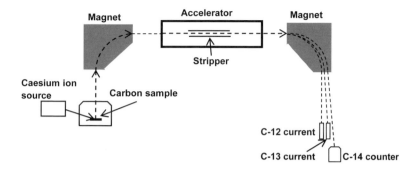

Figure 12.4   An accelerator mass spectrometer.

12.5 The Accelerator Mass Spectrometry (AMS) Method

The rate of beta-emission can be very low for older samples and to compensate for this either large samples or very long counting times must be used. The AMS method requires much smaller samples because it directly measures the number of atoms of each of the isotopes in the sample. The equipment used, illustrated in Figure 12.4, is similar to the mass spectrometer described in Section 8.8 but somewhat different to meet the particular needs of this kind of measurement.

Caesium ions are directed onto the carbon sample and negatively charged carbon atoms are spluttered from the surface and sent along a vacuum tube to the first magnet. This bends the beam of ions and only allows the carbon isotopes to pass through, any others present being deflected either more or less, so they do not pass out of the magnet. The beam then enters an accelerator which energizes the carbon ions. Within the accelerator they pass through a *stripper*, either a gas or a thin carbon foil, which strips electrons off them and they emerge with a positive change equal in magnitude to three electron charges. A second magnet separates the beam of carbon atoms into separate streams for the three isotopes. The C-14 stream enters a counter that directly counts the number of atoms arriving. The C-12 and C-13 streams are too intense to count individual atoms so instead, they enter detectors that measure the total charge they carry, from which the number of contributing atoms can be deduced.

The AMS requires carbon samples in the range of milligrams and, since even in such small samples there are large numbers of atoms, it gives the ages of older specimens with a much smaller standard deviation. However, it is still subject to the problem of the variability of the [C-14]/[C-12] ratio, a problem that we now address.

12.6 The Role of Dendrochronology and Similar Techniques

Dendrochronology is the science of dating based on the presence of tree rings. Under normal circumstances, the passage of one year of seasonal change will be marked by the addition of a new ring, visible in the cross section of a tree trunk (Figure 12.5). Each year a new ring grows on the outside, next to the bark, and the age of the tree is clearly indicated by counting the rings from the centre. The [C-14]/[C-12] ratio in each ring will be that at the time the ring formed, reduced by decay corresponding to its age.

The width of each ring reflects the growing conditions during the year it was produced, so the pattern of rings gives a climatic record from the time the tree came into being. A warm and wet year gives a wide ring, a dry year a narrow one. Within any restricted region of the world experiencing the same climatic conditions, the sequences of wider and thinner rings will be the same for different trees in that region — it has all the characteristics of a fingerprint. Tree rings give an absolute digital

Figure 12.5   Tree rings.

record of time — one ring, one year. In most regions of the world there will be old trees going back many centuries or even millennia and the tree-ring record from these, taken by removing cores without killing the tree, gives a standard continuous pattern of rings covering the period of the life of the tree. This can be used to date the wood used in the construction of old buildings. By matching the tree-ring pattern in the wood with the standard pattern, a date can be assigned to the wood corresponding to its outermost ring and, assuming that the wood was used shortly after the tree was felled, the building cannot be any younger than that; if the wood includes the outer region of the tree, corresponding to the latest date of ring formation, as is true for some roof trusses, then, again assuming that the wood was used shortly after the tree was felled, the date of the building may be estimated quite closely.

There are some extremely ancient trees in existence but not all contain tree-ring information. Their age is in the root system which throws up new trunks from time to time, but the ages of the trunks might be only a few hundred years. The oldest trunks from which ages may be deduced are those of bristlecone pines in California (Figure 12.6), the oldest being 5,064 years old. By matching tree-ring

Figure 12.6   An ancient bristlecone pine.

Figure 12.7   An outcrop of Pleistocene epoch varves in Ontario.

sequences with some dead specimens of bristlecone pine, dendritic sequences back to about 10,000 years can be established. Measurement of the [C-14]/[C-12] ratios in the rings and knowing their ages enables the original ratios to be determined and these show that there have been variations of the ratio with time.

Unrelated to tree-rings but of a similar character are *varves*, sedimentation layers in glacial lakes. Each spring and summer the glacier melts and the melt-water flows into the lake. The heavier material carried by the water is deposited even during the water flow but the lighter material is held in suspension at this time. When the water flow ceases in the autumn and winter, the lighter material is then able to settle. Each year is marked by layers of coarse and fine deposited material and contained in each layer is organic debris from plants. Sometimes the beds of ancient glacial lakes and cross-sections of the varves are exposed (Figure 12.7). Varve sequences are also available from cores taken from existing glacial lakes. As is the case for tree rings, the thickness and appearance of successive varves form a characteristic pattern and can be compared to link varves from different sites to give an extended chronological sequence, now going back more than 50,000 years. Information from varves supplements the information from tree rings and also gives information about the variation of climatic conditions, both from the thicknesses of the varves and the plant types they contain.

Another type of observation that can be used to extend knowledge about the [C-14]/[C-12] variation with time comes from corals,

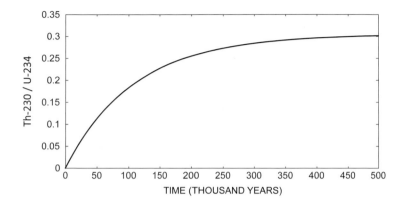

Figure 12.8   The variation of the ratio [Th-230]/[U-234] with time.

organisms in the sea living to a great age, with part of their structure in the form of calcium carbonate, an inanimate material with dissolved carbon dioxide as its source. The inanimate material also includes uranium and the decay chain of interest is that beginning with U-238, with half-life 4.47×10^9 years, the next step being U-234, with half-life 2.455×10^5 years, followed by the thorium isotope Th-230, with half-life 7.538×10^4 years. Uranium compounds are soluble but under normal conditions thorium compounds are not, so any thorium present must come about as a consequence of uranium decay. When the uranium first enters the coral, the ratio [U-234]/[U-238] is 5×10^{-5} and no thorium is present. As time passes, the ratio [Th-230]/[U-234] changes, as shown in Figure 12.8, and the ratio gives the age of the coral. Using this age, the initial ratio [C-14]/[C-12] can be found when the calcium carbonate first formed. Because the ratio [Th-230]/[U-234] saturates, tending towards a constant value, age can only be determined with any precision for about 250,000 years — well beyond the time for which C-14 can be detected.

Yet another indicator that can be used to investigate climatic changes in the past and also reveal changes in the [C-14]/[C-12] ratio is inclusions of carbon dioxide gas in Antarctic ice cores. A three kilometre core, formed by annually fluctuating snowfalls compressed into ice by the weight of subsequent falls, enables variations of climate to be followed for more than 800,000 years and the variation of C-14

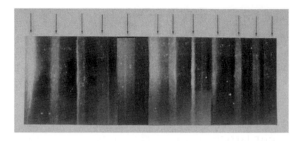

Figure 12.9 Part of an Antarctic ice core showing deposits over a 12 year period.

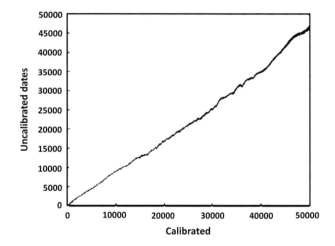

Figure 12.10 The correction curve giving true (calibrated dates) age against age based on the 1950 [C-14]/[C-12] ratio (uncalibrated dates).

to be found for the period of interest for archaeology. The section of a core displayed in Figure 12.9 shows 12 annual layers; they vary in width and appearance depending on the climatic conditions of the year the snow fell.

Using information from tree rings, varves, corals and ice cores, it has been possible to construct the variation of the [C-14]/[C-12] ratio over the period for which radiocarbon dating is applicable. This is then transformed to give the age-correction curve shown in Figure 12.10, which converts the age found assuming a constant [C-14]/[C-12] ratio at the 1950 value, to the true age.

12.7 Thermoluminescence Dating

Within a perfect crystalline solid, with atoms arranged on a regular lattice, the electric field is also regular, following a perfect periodic pattern. The atomic electrons are disturbed from the positions they would have in an isolated atom because they are being shared with, received from, or donated to, other atoms to bind atoms together to form the coherent lattice. If the perfect lattice is exposed to radiation (or particles) that give an electron enough energy to break free of its host nucleus, then the electron will move around the lattice, colliding with nuclei and other electrons until eventually it finds a nucleus with a missing electron and becomes a bound electron again. In this case, the net effect of the radiation is to confer energy of motion to the constituent atoms of the lattice, equivalent to increasing its temperature.

However, extensive perfect lattices are the exception rather than the rule and most lattice structures will contain *defects* — dislocations, missing atoms or extra atoms squeezed in. A two-dimensional representation of these three kinds of defect are shown in Figure 12.11.

The effect of a defect is locally to distort the electric field, producing bumps and troughs, above and below the values of the periodic variation due to a perfect lattice. This is illustrated in one dimension in Figure 12.12. Now an electron given energy to move through the lattice will occasionally reach a bump that, according to classical physics, it cannot pass through because its energy of motion is insufficient. However, electrons obey the physics of quantum mechanics and there is a small probability that it *can* penetrate the barrier, a phenomenon

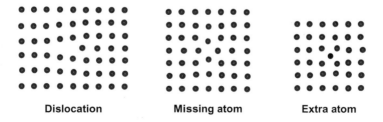

Dislocation   Missing atom   Extra atom

Figure 12.11   A representation of three kinds of defect.

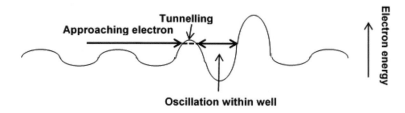

Figure 12.12   An electron becoming trapped in a potential well.

known as *tunnelling*. In the case shown in Figure 12.12, it is now in a well and it will bounce to and fro as shown. Every time it hits a wall of the well, there is a small probability that tunnelling will take place again and that it will escape, but it may remain in the well for a very long time — perhaps hundreds or thousands of years. The greater the total radiation dose — which depends on the product of the intensity of the radiation and the period of exposure — the greater the number of these trapped electrons in the lattice at any time.

When the lattice material is either heated or exposed to a powerful light source, the trapped electrons acquire enough energy to escape from the potential wells and continue moving through the lattice, eventually to combine with a nucleus that is deficient in its electron complement. When it does so, the electron releases energy in the form of an ultraviolet photon — *thermoluminescence* — and the total energy of ultraviolet light emitted, E_t, is a measure of the number of electrons that were trapped and hence of the radiation dose it received since the lattice last existed without any trapped electrons.

Although there will be some radiation falling on the lattice material from outside due to cosmic rays and natural terrestrial radioactivity, the bulk of the radiation will be from traces of radioactive elements within the material itself, such as uranium, thorium or the isotope potassium-40. Thermoluminescence dating is most often used to date pottery or ancient hearth material. When the pottery is fired or the hearth is used, all trapped electrons are released and this sets the time to zero for trapping electrons in the cooled material. The radioactivity is measured in a fixed amount of the material in the order of a few grams and the contribution from cosmic rays can be added to this. This gives the dose per unit of time, say r, the dose per year, during the period since the sample cooled. Next the material is calibrated by exposing it to a very strong known dose of radiation D and the intensity of the ultraviolet light emitted, E_c, is measured. The energy emitted is proportional to the total dose of radiation, so if the age of the sample is T then

$$\frac{rT}{D} = \frac{E_t}{E_c}$$

or

$$T = \frac{E_t D}{E_c r}. \tag{12.9}$$

The lower limit of time for dating artefacts is set by the minimum amount of emitted light that can be measured and the upper limit by saturation of the electron traps. This depends on the nature of the specimen but is about 100–200,000 years. The method is not very precise, with an expected error in the order of 15%, but it can be used for specimens without carbon content, for which radiocarbon dating is not possible. It is particularly valuable in the detection of fake antique pottery, although the fakers can artificially irradiate their wares to prevent detection by a thermoluminescence test.

Index